2nd International Conference on Composite Materials Science and Technology (ICCMST 2019)

IOP Conference Series: Materials Science and Engineering
Volume 600

Tokyo, Japan
22 – 25 May 2019

ISBN: 978-1-5108-9478-5
ISSN: 1757-8981

Printed from e-media with permission by:

Curran Associates, Inc.
57 Morehouse Lane
Red Hook, NY 12571

Some format issues inherent in the e-media version may also appear in this print version.

Institute of Physics
Dirac House, Temple Back
Bristol BS1 6BE UK

Phone: 44 1 17 929 7481
Fax: 44 1 17 920 0979

techtracking@iop.org

Additional copies of this publication are available from:

Curran Associates, Inc.
57 Morehouse Lane
Red Hook, NY 12571 USA
Phone: 845-758-0400
Fax: 845-758-2633
Email: curran@proceedings.com
Web: www.proceedings.com

Table of contents

Volume 600

2019 The 2nd International Conference on Composite Materials Science and Technology

22–25 May 2019, Tokyo University of Science

Accepted papers received: 31 July 2019
Published online: 20 August 2019

Preface

Papers

Composite Preparation

Property Analysis of Composite

IOP Conf. Series: Materials Science and Engineering **600** (2019) 011001 doi:10.1088/1757-899X/600/1/011001

PREFACE

The 2019 International Conference on Composite Materials Science and Technology (ICCMST 2019) was appeared in Tokyo, Japan, from 22 May to 25 May, 2019. On behalf of the organizing committee of this conference, it is my great pleasure and honor to introduce the proceedings of ICCMST 2019.

The proceedings include a variety of advanced research topics which were presented in the conference. Full papers submissions were blind peer-reviewed by at least two experts from the International Program Committee in order to ensure high quality standards of research. The selection process followed strict criteria and consequently only 27 full papers and 35 abstracts were selected and accepted for presentation on conference.

In the conference, six parallel oral sessions and poster session were well organized with scientists from 23 countries. Active discussions about new scientific data, practical techniques, and future perspectives in composite materials and nanomaterials, chemical and materials engineering, nanotechnology etc. In the evening, they joined to the conference dinner held at a buffet restaurant of the Tokyo Dome Hotel.

We certainly appreciate the efforts and contribution from the program chairs, organization staff and the members of the program committee for their work. Thus, we wish to express our gratitude to all those who have contributed to the success of ICCMST 2019. We sincerely hope that all participants and guests benefit significantly from the published proceedings.

At last, we hope all of you will again get together at the next ICCMST, to have intelligent and exciting, academic and practical discussions which will be rewarding and most enjoyable experience.

With our warmest regards,

Prof. Kazuo Umemura, Tokyo University of Science,
Japan

22 May to 25 may, 2019
Tokyo, Japan

IOP Conf. Series: Materials Science and Engineering **600** (2019) 011002 doi:10.1088/1757-899X/600/1/011002

Conference Chairs

Prof. Kazuo Umemura, Tokyo University of Science, Japan
Prof. Takashi Amemiya, Nippon Institute of Technology, Japan
Prof. Masayuki Morimoto, Tokai University, Japan
Prof. Ab Halim Bin Abu Bakar, University of Malaya, Malaysia

Program Chairs

Prof. Henry Hu, University of Windsor, Canada
Prof. Toshiyuki Yasuhara, Nippon Institute of Technology, Japan
Prof. Kei Eguchi, Fukuoka Institute of Technology, Japan
Prof. Seung-Hoon Yoo, Seoul National University of Science and Technology, Korea

Technical Committee

Prof. Pranut Potiyaraj, Chulalongkorn University, Thailand
Prof. Jun-Dr Hwang, National Chiayi University, Taiwan
Prof. Bolyu Xiao, Institute of Metal Research, Chinese Academy of Sciences, China
Prof. Eduardo Sanchez, University of Seville, Spain
Prof. Ricardo Nuno Madeira Soares Branco, Universidade de Coimbra, Portugal
Prof. Camelia CERBU, Transilvania University of Brasov, Romania
Prof. Adrian VOLCEANOV, Bucharest Polytechnic University, Romania
Prof. Konstantinos Beltsios, University of Ioannina, Greece
Prof. Jose Xavier, Universidade do Porto, Portugal
Prof. Kanji Ono, University of California, Los Angeles, USA
Prof. Kyong-Yop Rhee, Kyung Hee University, Korea
Prof. Enda McGlynn, Dublin City University, Ireland
Prof. Leif Asp, Chalmers University of Technology, Sweden
Prof. Bingqing Wei, University of Delaware, USA
Prof. Volkan Gunay, Yeditepe University, Turkey
Prof. Metin Aydogdu, Trakya University, Turkey
Prof. Juan Du, SIOFM, Chinese Academy of Sciences, China
Prof. Binghui Luo, Central South University, China
Prof. Gaohui Wu, Harbin Institute of Technology, China
Prof. Wu Sujun, Beihang University, China
Prof. Wu Tzong-Ming, National Chung Hsing University, Taiwan
Prof. Guohua Chen, Huaqiao University, China
Prof. Yuan Hua, Tongji University, China
Prof. Zhen Yudong, University of Science and Technology Beijing, China

Peer review statement

All papers published in this volume of *IOP Conference Series: Materials Science and Engineering* have been peer reviewed through processes administered by the proceedings Editors. Reviews were conducted by expert referees to the professional and scientific standards expected of a proceedings journal published by IOP Publishing.

IOP Conf. Series: Materials Science and Engineering **600** (2019) 012001 doi:10.1088/1757-899X/600/1/012001

Graphene-enhanced ABS for FDM 3D printing: effects of masterbatch preparation techniques

Kittinon Sakunphokesup[1, a*], Poomipat Kongkrengkri[1,b], Aphiwat Pongwisuthiruchte[1,c], Chuanchom Aumnate[2,d] and Pranut Potiyaraj[1,2,e]

[1]Department of Materials Science, Faculty of Science, Chulalongkorn University, Thailand, [2]Metallurgy and Materials Science Research Institute, Chulalongkorn University, Thailand

[a]sometime28988@hotmail.co.th; [b] poouza555@gmail.com; [c]riceoat@gmail.com; dchuanchom.a@chula.ac.th; [e]pranut.p@chula.ac.th

Abstract. 3D printing technologies have received a lot of attentions because of the high flexibility of the process as well as the capability of producing complex shaped and high detailed artifacts with ease. This study focuses on the development of acrylonitrile butadiene styrene (ABS)/graphene composite materials, which can be used as a filament for the melt extrusion 3D printing or FDM 3D printing technique. Since the dispersity of graphene in the polymer matrix is the key parameter of controlling the properties of the composites, the ABS/graphene masterbatches with high concentration of graphene were produced via three techniques: dry mixing; solvent mixing and coagulation techniques. The obtained high concentration ABS/graphene masterbatches were then mixed with pure ABS and melt extruded to produce ABS/graphene composites filament with 1 wt% of graphene. The ABS/graphene composites preparing via solvent mixing and coagulation were found successfully 3D printed. The dry mixing method brought about the aggregation of graphene particle and failed the print.

Keywords: Acrylonitrile butadiene styrene; Graphene; 3D Printing; Nanocomposite; Fused Deposition Modeling

1. Introduction

3D printing technologies have gained a huge popularity due to the capability of making a customized part easily without a requirement of large-scale production. Among various 3D printing techniques, the fused deposition molding (FDM) technology is currently the most widely available 3D printing platforms. This technique uses a tiny die with a heater to melt thermoplastic materials in the form of a continuous filament. The melted materials were then extruded and building up the 3D object on the platform. However, the mechanical properties of FDM 3D printed parts are different from those fabricated by the conventional processes, especially the injection molding process [1]. ABS is one of the most common thermoplastics used to produce filament for 3D printing. Graphene is allotropy of carbon crystal similar to graphite diamonds or carbon nanotubes. Graphene is very strong, lightweight, almost transparent and has a good conductivity both heat and electricity. Fabricating ABS/graphene composites will potentially enlarge the applications and functionalities of the 3D printed products [2].

IOP Conf. Series: Materials Science and Engineering **600** (2019) 012001 doi:10.1088/1757-899X/600/1/012001

However, the effects of the preparation method used to prepare the ABS/graphene composites are rarely demonstrated. This study aims to investigate the optimal method for mixing ABS and graphene for fabricating good-quality and functional filaments for the FDM 3D printing process.

2. Materials and method

2.1. Materials

The graphene used in this study was prepared from graphite using the improved Hummer's method [3]. The polymer matrix used in this work was ABS (AP102, IRPC Public Company).

2.2. ABS/graphene composite filaments preparation

The ABS/graphene composite filaments were prepared by the melt extrusion technique. The masterbatches were prepared via three methods. First is the coagulation method in which the obtained graphene was first dispersed in DMF and mixed with the ABS solution in DMF. The concentration of graphene in the ABS solution was set to 10wt%. The ABS/graphene composite solution was then precipitated in methanol. The obtained precipitated ABS/graphene was dries and mixed with virgin ABS to further produce the 1.75 mm diameter ABS/graphene filament by the melt extrusion technique. The obtained filament was called ABS/G1A. The second method is the solvent mixing method. Here, the ABS/graphene solution was prepared using the same method used for the coagulation sample. Instead of precipitation in methanol, DMF was evaporated. The obtained ABS/graphene composite after solvent evaporated was then mixed with virgin ABS and melt extruded to produce the ABS/graphene filament, which further called as ABS/G1B. The last method is the dry mixing method, which the dry graphene powder was mixed with virgin ABS and melts extruded to produce the filament. The filament prepared via the dry mixing method was further called ABS/G1C.

2.3. Fabrication of 3D printed samples

The tensile testing specimens (Type V, according to the ASTM D638-14) [4] were 3D printed using the WANHAO Duplicator 6 FDM 3D printer. Each sample was printed individually at the center of the printing bed by using the parameters given in Table 1.

Table 1. 3D printing parameters.

Parameters	
Nozzle diameter [mm]	0.4
Layer Height [mm]	0.1
Infill [%]	30
Printing Temperature [^{o} C]	230
Build Plate Temperature [^{o} C]	80
Printing Speed [mm/s]	25

2.4. Characterization

The graphite oxide and the obtained graphene were characterized using an X-ray diffractometer (Bruker AXS Model D8 Discover). The graphene sample was scanned over the range of $2\theta = 2$ - $50°$ and measurements were recorded at every 0.02° interval.

The decomposition of pure ABS, ABS/G1A, ABS/G1B, and ABS/G1C were analyzed using the TGA model TG 209 F3 Tarsus®. The samples were heated from 50-700 degrees Celsius with 10K/min ramping step under the nitrogen atmosphere.

The melt flow rate (MFR) of ABS/graphene composites; ABS/G1A, ABS/G1B and ABS/G1C were measured according to ASTM D1238 standard, at a temperature of 240 °C with 2.16 kg load applied [5].

The tensile properties of all 3D printed specimens were measured according to ASTM-D638 standard using a Universal Testing Machine equipped with a 5 KN load cell at a crosshead speed of 5 mm/min.

3. Results and discussion

The crystal structures of graphite, graphite oxide and the synthesized graphene were analyzed by XRD technique and presented in Fig.1. The characteristic peaks of graphite was observed at $2\theta = 26.46°$. The peak of graphite oxide was shifted to $2\theta = 11.26°$. The distance between the carbon atom planes of graphite increases from the oxidation reaction to graphite oxide as can be seen from the broader peak, representing the increase in the interlayer spacing. After the reduction step, the sharp peak of graphite oxide disappeared and the broad peak showed up at $2\theta = 25.42°$. The broad and low intensity of the peak indicated the disorder of the graphene particle structure.

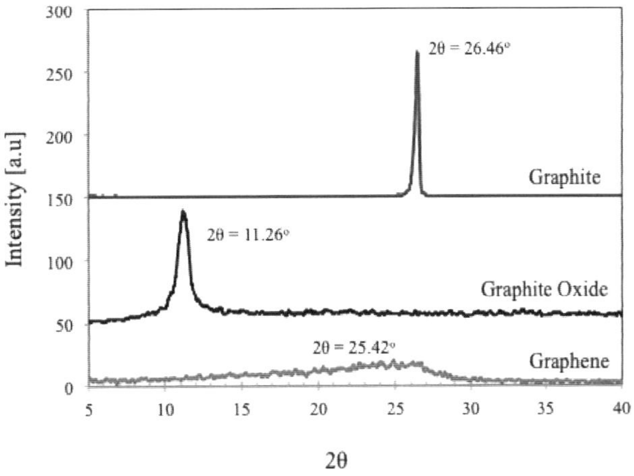

Figure 1. X-ray diffraction graph of graphite, graphite oxide and synthesized graphene.

Fig.2 shows the TGA thermogram of pure ABS, ABS/G1A, ABS/G1B, and ABS/G1C composite filaments. The ABS/graphene composite materials showed slightly higher initial decomposition temperatures; compared to the pure ABS. However, the decomposition of the ABS/graphene composites prepared from the different three methods appeared at almost the same temperature. Thus, it could be concluded that the different methods used to prepare the ABS/graphene composite in this work had no significant effect on the thermal stability of the obtained composites.

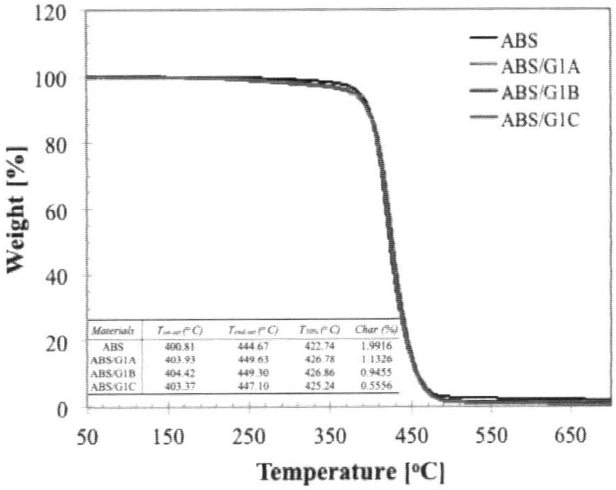

Materials	$T_{on\text{-}set}$ (°C)	$T_{end\text{-}set}$ (°C)	$T_{50\%}$ (°C)	Char (%)
ABS	400.81	444.67	422.74	1.9916
ABS/G1A	403.93	449.63	426.78	1.1326
ABS/G1B	404.42	449.30	426.86	0.9455
ABS/G1C	403.37	447.10	425.24	0.5556

Figure 2. TGA thermogram of ABS, ABS/G1A, ABS/G1B and ABS/G1C.

Fig. 3 shows the MFR values of the three ABS/graphene composites as well as the pure ABS. The MFR values of pure ABS, ABS/G1A, ABS/G1B and ABS/G1C are 6.7± 0.8, 6.*8* ± 1.4, 6.5 ± 1.0 and 7.0 ± 0.1 g/10min, respectively. It was found that the MFR values slightly increased after the graphene was incorporated. However, the ABS/G1A showed the largest extent of deviation, which may be attribute to the inhomogeneous of the material.
Fig. 4 shows the tensile properties of ABS, ABS/G1A, ABS/G1B and ABS/A1C composites. Incorporating 1wt% of graphene significantly enhanced the tensile properties of the pure ABS. The coagulation (ABS/G1A) and dry mixing methods (ABS/G1C) gave a comparable Young's modulus values, however, the composite prepared by the coagulation method delivered the better improvement in the tensile strength and elongation at break. The solvent mixing method (ABS/G1B) offered the best improvement in the tensile strength, Young's modulus as well as the Elongation at break.

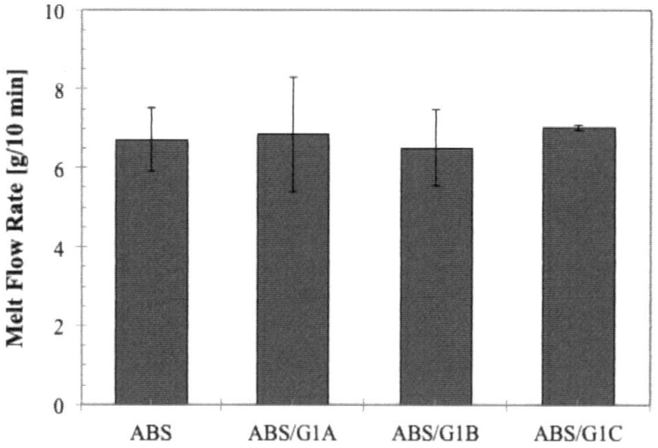

Figure 3. Melt flow rate (MFR) of ABS and ABS/graphene composites; ABS/G1A, ABS/G1B and ABS/G1C.

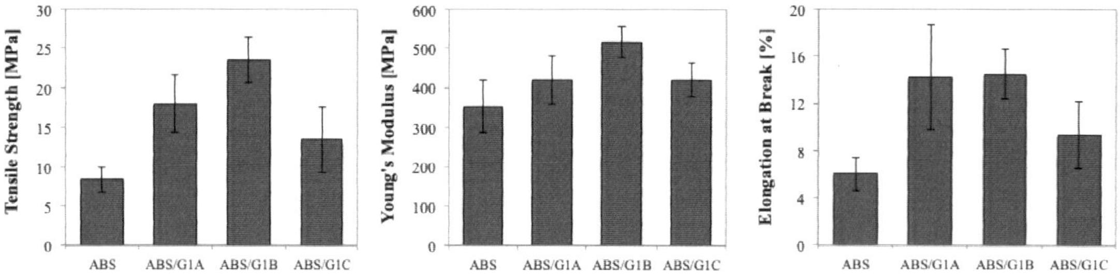

Figure 4. Tensile properties of 3D printed specimens from ABS, ABS/G1A, ABS/G1B and ABS/G1C.

4. Conclusion

Graphene was successfully used to enhance the mechanical performance of ABS as evidenced from the improvement in tensile properties. Among three methods used to prepare the ABS/graphene masterbatchess in this study, no significant different in the flowability and thermal stability were be observed. The ABS/graphene composite preparing by the solvent mixing method (ABS/G1B) exhibited the best improvement in the tensile strength, elongation at break and Young's modulus. Therefore, the solvent mixing method seems to be the most promising method and will be used to prepare the ABS/graphene composites for further study.

References

[1] M. Dawoud, I. Taha, S.J. Ebeid: Journal of Manufacturing Processes Vol. 21 (2016), p. 39-45.

[2] C. Aumnate, A. Pongwisuthiruchte, P. Pattananuwat and P. Potiyaraj: Advances in Materials Science and Engineering Vol. 2018 (2018), p. 1-9. https://doi.org/10.1155/2018/2830437
[3] Lasprilla, A.J.R., G.A.R. Martinez, and B. Hoss, Synthesis and characterization of poly (lactic acid) for use in biomedical field. Chem. Eng, 2011. 24: p. 985-990.
[4] ASTM International: Standard Test Method for Tensile Properties of Plastics (D638-14). ASTM Int., West Conshonhocken, PA, USA, (2014).
[5] ASTM International: ASTM D1238-13, ASTM Int., West Conshonhocken, PA, USA, (2013).

IOP Conf. Series: Materials Science and Engineering **600** (2019) 012002 doi:10.1088/1757-899X/600/1/012002

Preparation of polypyrrole coated zinc anode electrode for inhibition corrosion of secondary zinc ion battery

W Nuanwat[1,3] and P Pattananuwat[1, 2, 3,*]

[1]Department of Materials Science, Faculty of Science, Chulalongkorn University, Bangkok, 10330, Thailand
[2]Research Unit of Advanced Materials for Energy Storage, Chulalongkorn University, Bangkok, 10330, Thailand
[3]Center of Excellence on Petrochemical and Materials Technology, Chulalongkorn University, Bangkok, 10330, Thailand

E-mail: *prasit.pat@chula.ac.th

Abstract. In this work, polypyrrole coated onto zinc surface anode electrode was successfully prepared via cyclic voltammetry deposition. The different PPy thickness layer on zinc electrode is controlled by the different numbers of deposited cycles at 3, 5, 7 and 10 cycles. The surface morphology development of polypyrrole nano/micro structure covering on zinc surface was examined by Scanning Electron Microscopy-Energy Dispersive X-Ray Spectroscopy. The electrochemical activity of polypyrrole coated zinc electrode is characterized by linear polarization, cyclic voltammetry and galvanic static charge-discharge, indicating that the anti-corrosion PPy surface covering on Zn electrode can yield the highest value of corrosion inhibition efficiency of 67% and serve as high redox layer to improve the charge performance.

1. Introduction

Recently, the development of anode materials for metal ion secondary batteries (such as lithium-ion, aluminium-ion, potassium-ion, sodium-ion and zinc-ion) has attracted increasing attention. Among of them, lithium-ion are most widely materials used owning to its features, possessing high voltage, high energy density and long-life cycle. However, the price of lithium tends to increase due to the plentiful consumption of lithium in future. Thus, the study the sustainable materials for secondary batteries is necessary. Zinc (Zn) is one of the most commonly used as promising anode materials for secondary battery due to its high theoretical density energy, abundant resource and environmentally begin materials that are used [1]. Nevertheless, zinc-based anode electrode reveals some issues, including the corrosion and dendrite formation of anode electrodes during the charge-discharge process [2]. To overcome those problems, recently, conducting polymer such a polypyrrole which can form tightly on nanoparticle surface is considered as the shield to minimize the corrosion reaction of zinc electrodes and prevent the growth of zinc dendrite formation [3]. Several reports revealed the same results of the using conductive coated onto zinc particles which is not only to decrease corrosion rate but also can decrease a self-discharge behaviour enhancing the performance of zinc ion electrode [4]. However, to achieve the high efficiency of the corrosion suppression during discharge process, the design homogenous uniform layer of coating materials on zinc surface is priority to concern. Polypyrrole is one of interesting conductive polymer with high redox reaction. Its characteristic features possess the flexible and the charge molecule offering high compatibility with metal surface. But, the use of polypyrrole as coating materials for zinc surface still has not been widely addressed. Therefore, in this work, the incorporation of polypyrrole as a conductive polymer layer coated zinc surface is purposed as shielding layer for suppressing the corrosion rate. The different amounts of polypyrrole coated zinc surface on corrosion rate was also studied simultaneously with the electrochemical performance for zinc-manganese rechargeable battery.

2. Experimental section

2.1. Preparation of bare zinc and polypyrrole coated zinc anode electrode

The bare zinc and polypyrrole coated zinc electrode were prepared by electrochemical deposition. Firstly, the electroplating technique, configurated with parallel electrode consisting of nickel foam as cathode and zinc plate as anode, was used to deposit zinc layer on nickel foam with supporting electrolyte of 0.6 M zinc sulfate aqueous solution. The current density at 65 mA cm^{-2} was applied for 10 minutes using potenstat/galvanostat (PGSTAT204) [5]. The obtained zinc electrode was wash with distilled water and dried overnight. For polypyrrole coated zinc electrode, the cyclic voltammetry technique, configurated with conventional three electrode systems consisting of the as-prepared zinc electrode as working electrode, Ag/AgCl as reference electrode and platinum as counter electrode, was used to deposit the polypyrrole on zinc surface electrode. The voltage between 0 to 1 V at scan rate 50 mV sec^{-1} with 3 cycle depositions was applied with supporting electrolyte of 0.1 M pyrrole aqueous solution at pH 3. In this present work, the different amounts of polypyrrole coated zinc electrode were prepared by modulating the number of deposited cycles at 0, 3, 5, 7 and 10 cycles (denoted as Zn, Zn/PPy-3, Zn/PPy-5, Zn/PPy-7 and Zn/PPy-10, respectively).

2.2. Preparation of δ-MnO_2 cathode electrode

δ-MnO_2 was synthesised by hydrothermal method as described elsewhere [6]. The cathode electrode was prepared by pasting the mixtures of 80 %wt. of δ-MnO_2, 15 %wt. of conductive carbon and 5 %wt. of polytetrafluoroethylene in water/ethanol solution. The mixture slurry was pasted on Ni foam, pressed and placed in oven at 60 °C for 3 h. Finally, the as-prepared electrode was cut in size 1 cm x 2 cm.

2.3. Characterization

The characteristic functional groups of the samples were examined by Fourier transform infrared spectroscopy (FT-IR) with a Thermo scientific Nicolet 6700 (Tokyo, Japan). Polypyrrole coated zinc electrode was dried and coated with 10 nm of gold before being observed the surfaces using scanning electron microscopy (SEM: JSM-6480LV) at an acceleration voltage of 10 keV.

2.4. Electrochemical measurements

To evaluate the corrosion of the as-prepared Zn coated PPy, linear polarization technique was applied to conventional three-electrode configuration with the as-prepared samples as working electrode, platinum plate as the counter electrode and Ag/AgCl as the reference electrode by scanning between -1.0 and -0.9 V vs E_{we} its open circuit voltage (OCV) at the rate of 0.5 mV/s. To evaluate electrochemical performance, cyclic voltammetry was determined with two electrodes consisting of the as-prepared samples as working electrode and δ-MnO_2 as the counter electrode, employing in 50 ml of 2 M zinc sulfate between 1 to 2 V at 5 mV/sec. The galvanostatic charge/discharge technique was used to evaluate the energy capacity of Zn-Mn battery. The positive electrode was the δ-MnO_2 with a much higher capacity than that of the zinc electrode three times. The negative electrode was the zinc electrode. The cells were charged 4C for 20 min and discharged at 1C rate down to 1.3 V

3. Results and discussion

3.1. Morphology study

In this work, polypyrrole is coated on zinc surface by cyclic voltammetry deposition. Representative surface morphologies of bare zinc and polypyrrole coated zinc at different number of deposited cycles are illustrated in Fig.1. SEM evidence in Fig.1a reveals the stacking of flat plate-like structure of zinc. Obviously, the remarkable difference of the morphology in each PPy coated zinc samples can be easily recognized, demonstrating that the increase in dense growth of ppy covering on zinc surface with increase the number of cycle deposition (seen Fig. 1(b), (c), (d) and (e), respectively). Additionally, SEM-EDS evidence in Fig.1(f) reveals the nitrogen signal on bulk surface, confirming that the coating layer on zinc surface is polypyrrole.

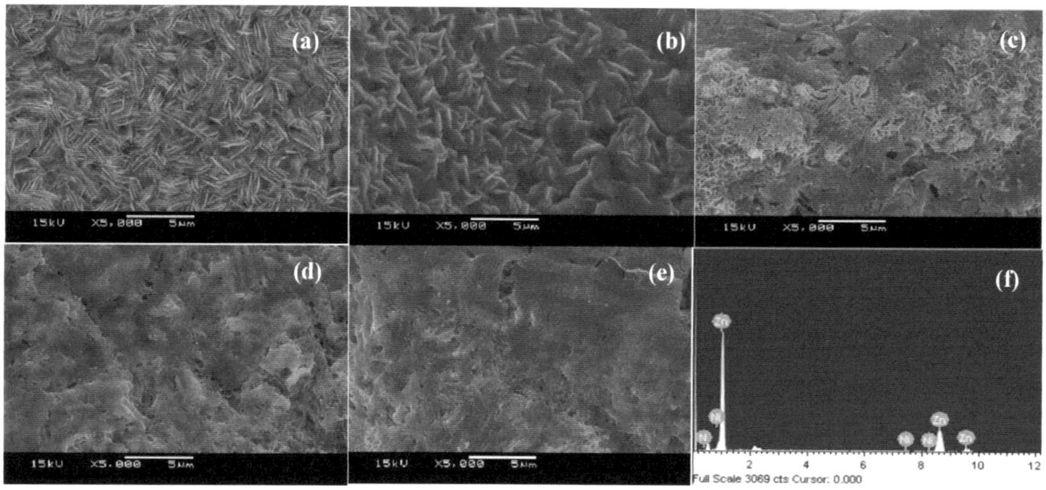

Figure 1. SEM images of (a) Zn (b) Zn/PPy-3, (c) Zn/PPy-5, (d) Zn/PPy-7, (e) Zn/PPy-10 and (f) SEM-EDS of Zn/PPy-3

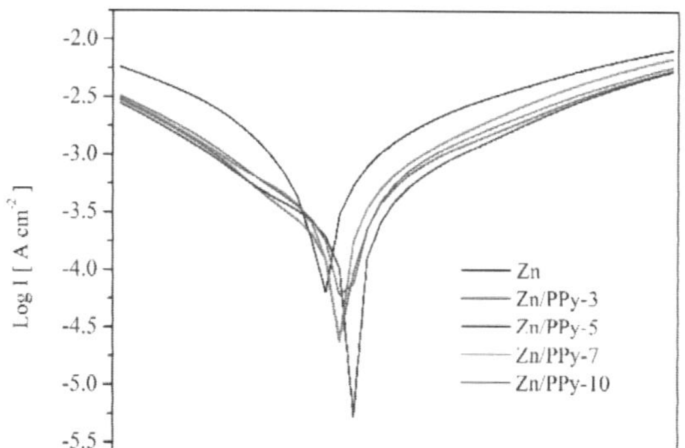

Figure 2. Tafel polarization curves for zinc and polypyrrole coated zinc electrodes

3.2. Electrochemical properties

The linear polarization curves of the bare zinc and polypyrrole coated zinc electrodes are illustrated in Fig. 2. The calculated potential and current corrosion by Tafel fit are tabulated in table 1. As seen from table 1, obviously, the bare zinc electrode possesses the highest corrosion currents at 483.17 $\mu A/cm^2$, while all polypyrrole coated zinc electrode reveal the lower corrosion currents than that of zinc, implying that the anti-corrosion performance of polypyrrole coated zinc is superior to zinc. As SEM evident support, PPy coated zinc surface can act as shielding effect to decrease the direct contact between zinc and electrolyte leading to slow down the zinc corrosion process. It can be concluded that polypyrrole coating on zinc surface can suppress the corrosion of zinc electrode. Zn/PPy-5 shows the lowest corrosion currents at 159.38 μA/cm^2 with the highest values of corrosion inhibition efficiency of 67.01%. The more coating surface of Zn/PPy-7 and Zn/PPy-10 electrode reveals the slight decrease in its corrosion protection. This result is ascribed to the fact that the highly dense PPy surface can limit the ion mobility and the reaction channels by trapping inside PPy layer, leading to accumulate the charge on its surface inducing in the polarization process.

Table 1. Corrosion parameters for zinc and polypyrrole coated zinc electrodes

Materials	E_{corr} (V)	I_{corr} (μA/cm^2)	η (%)
Zn	-0.96165	483.17	-
Zn/PPy-3	-0.95805	160.13	66.86
Zn/PPy-5	-0.95624	159.38	67.01
Zn/PPy-7	-0.95822	179.87	62.77
Zn/PPy-10	-0.95895	244.40	49.42

The electrochemical performance evaluated by cyclic voltammetry is displayed as seen in Fig. 3. All of polypyrrole coated zinc electrodes exhibit the enhancing in the CV area relative to bare Zn electrode, implying higher activity surface of PPy coated Zn. Galvanic static charge-discharge is used to determine the capacity of electrode at discharge rate of 1 C. As seen from Fig. 3(b), the discharge efficiency of polypyrrole coated zinc obviously reveal the better performance than that of bare zinc. The reason is that the conductivity of zinc electrode and redox reaction is improved by the polypyrrole layer. Zn/PPy-5 electrode reveals the highest capacity which is good agreement with SEM and linear polarization results. In addition, at high amount of polypyrrole coating, Zn/PPy-7 and Zn/PPy-10 exhibit the slight decrease in discharge efficiency. As evident support by SEM, Zn/PPy-7 and Zn/PPy-10 demonstrate the high thickness layer which can serve as trapping layer, causing the limitation of ion diffusion. The discharge products may retain in the electrode rather than dissolving into electrolyte leading to decrease in electrochemical capacity.

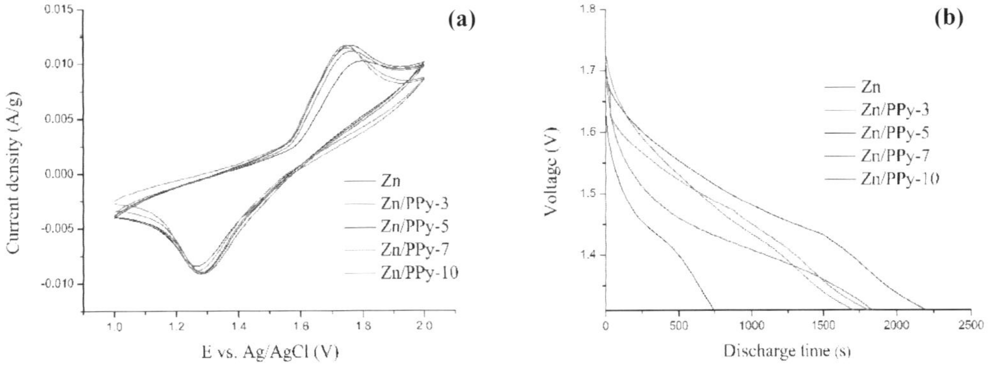

Figure 3. (a) Cyclic voltammograms of zinc and polypyrrole coated zinc electrodes at 5 mV/s and (b) Typical discharge curve of zinc and polypyrrole coated zinc electrodes at the rate of 1C

IOP Conf. Series: Materials Science and Engineering **600** (2019) 012002 doi:10.1088/1757-899X/600/1/012002

4. Conclusions

Polypyrrole coated zinc electrodes have been electrically synthesized via cyclic voltammetry deposition. The developed film thickness of PPy coated Zn depends on the number of deposited cycles. PPy coated Zn at 5 cycles reveals the most effective electrode for improving anti-corrosion with the highest value of corrosion inhibition efficiency of 67%. Evident results in CV and GCD reveal the better electrochemical capacity of PPy coated Zn electrode relative to zinc, especially at 5 cycles deposition. This polypyrrole coating zinc are considered as the high efficiency anode materials utilizing for zinc-ion batteries.

References

[1] McLarnon, Frank R and Cairns E J 1991 The secondary alkaline zinc electrode. Journal of the Electrochemical Society Vol. **138**, pp.645-656.

[2] Xu C, Li B, Du H and Kang F 2012 Energetic zinc ion chemistry: the rechargeable zinc ion battery. Angewandte Chemie International Edition Vol. **51**, pp.933-935.

[3] Lee S H, Yi C W and Kim K 2010 Characteristics and electrochemical performance of the TiO2-coated ZnO anode for Ni– Zn secondary batteries. The Journal of Physical Chemistry C Vol. **115**, pp.2572-2577.

[4] Parker J F, Chervin C N, Pala I R, Machler M, Burz M F, Long J W and Rolison D R 2017 Rechargeable nickel–3D zinc batteries: An energy-dense, safer alternative to lithium-ion. Science Vol. 356, pp.415-418.

[5] Sun K E, Hoang T K, Doan T N L, Yu Y, Zhu X, Tian Y and Chen P 2017 Suppression of dendrite formation and corrosion on zinc anode of secondary aqueous batteries. ACS applied materials & interfaces Vol. **9**, pp.9681-9687.

[6] Konarov A, Voronina N, Jo J H, Bakenov Z, Sun Y K and Myung S T 2018 Present and future perspective on electrode materials for rechargeable zinc-ion batteries. ACS Energy Letters Vol. **3**, pp.2620-2640.

IOP Conf. Series: Materials Science and Engineering **600** (2019) 012003 doi:10.1088/1757-899X/600/1/012003

Rice husk ash-geopolymer composite

Krissanapat Yomthong, Suthee Wattanasiriwech and Darunee Wattanasiriwech

Center of Innovative Materials for Sustainability,School of Science, Mae Fah Luang University, Chiangrai, Thailand

corresponding author: suthee@mfu.ac.th; darunee@mfu.ac.th

Abstract. This research aimed to enhance properties of fly ash based geopolymer through usage of three types of rice husk ash (RHA). The replacement in the content of 0%, 3%, 5% and 10% by weight was chosen. The RHA used in this study were obtained from (i) the rice drying-process (ii), (iii) in house by burning RH at 600°C and 700 °C. Liquid sodium silicate (LSS) and sodium hydroxide (SH) solution were used as activating agents. The weight ratio of LSS to SH was 1.0 and total liquid to solid was 0.6. The geopolymer prepared without RHA (the control) gave an average compressive strength of 35.91±3.30 MPa. All types of RHA replacement at 3% resulted in an improvement of compressive strength. Further increasing RHA content resulted in an adverse effect on workability and compressive strength of the final products. The highest compressive strength value of 42.52±4.86 MPa was obtained at 3% replacement using RHA burnt at 700 °C. The density results showed a decreased when incorporation of high content of RHA was beyond 5%.

1. Introduction

Approximately, one ton of carbon dioxide (CO_2) was released to the atmosphere per one ton of Portland cement (PC) production. This PC production also consumed a significant amount of natural resources and contributes to greenhouse problems [1] Finding a more environmental friendly material to replace PC is a critical challenge for scientist. Said material should have properties similar to PC, thus making geopolymer the most interesting material due to its likeness. Geopolymer can be synthesized from rich silica and alumina source materials. The rich silica and alumina source materials are usually comprised of fly ash (FA) from coal burning processes, thermally activated clay and slag from blast furnace burning process [2, 3]. The silica and alumina are activated by an alkaline activator that consists of alkali silicate and alkali hydroxide to produce aluminosilicate gel [4]. The curing temperature of geopolymer is at room temperature. The reaction between rich silica and alumina source materials and alkali activated solution is called geopolymerization. Geopolymers possess excellent mechanical properties, fire resistance, acid resistance and can be an alternative to PC [4, 5].

FA is a useful material for producing geopolymer because of its high alumina and silica contents. Use of FA as a precursor for geopolymer production could also offer benefits on waste management. The most widely used alkali activation solutions are sodium hydroxide (NaOH) and sodiumsilicate (Na_2SiO_3) because it is low cost and viability [6].

Attempts of improving geopolymer properties by adding nanomaterial showed that the mechanical properties could be enhanced [7]. Notwithstanding, nano-particle (SiO_2 and Al_2O_3) is expensive. An alternative material containing high amounts of silica is rice husk ash (RHA) so it is of interest to replace nano-silica. When rice husk (RH) was burned to eliminate carbonaceous compound at 600°C and 700°C, high purity silica could be obtained. This silica is porous and has high abundance of Si-OH group [8, 9].

The use of purified rice husk ash (RHA) containing a high amount of silica showed a positive effect on mechanical properties of geopolymer. However, preparation of high purity RHA was too costly to be commercially viable. Partially burned RHA, an abundant by-product from rice husk (RH) fired hot air generator, could be a more favorable choice.

This research thus aims to explore the use of partially burned RHA in geopolymer formation to improve the mechanical property. The replacement of FA with RHA in the content of 0%, 3%, 5% and 10% by weight were chosen. The amount lower than 3% did not give interesting result so the selection was limited to only greater amounts.

2. Experiment

FA was received from Lampang, Thailand. The activated alkali liquids used were 12M NaOH and commercial grade liquid sodium silicate (LSS). The chemical compositions of LSS was 16.24% Na_2O, 34.73% SiO_2 and 49.03% H_2O. The weight ratio of SS /SH used was 1.0 and ratio of total liquid to solid (L/S) used was 0.6. Three types of RHA were used in this experiment: RHAR, RHA600, RHA700. RHAR was received from a rice-drying process place without further modification. RHA600 and RHA700 was derived from in house burning of RH in a muffle furnace (Carbolite/ CWF13/13/2416; Scientific Promotion) at 600 °C and 700 °C for 4 h respectively. This condition was selected due to the highest amount of nanosilica that could be obtained [8]. FA was milled using ball milling for 10 min. RHAs were hand grounded with a mortar for 5 min. Particle sizes of FA, RHAR, RHA600 and RHA700 after milling and grinding were determined by a Particle Size Analyzer (Malvern Instrument/Masterzer 2000).

The LSS and SH were hand-mixed together for 2 min and used as a liquid solution. Subsequently, RHA was added into liquid solution at the content of 0%, 3%, 5% and 10% FA replacement and mixed for 2 min. FA was then added and mixed for 1 min. Finally, precursor paste was poured into PVC mold. The samples were cure at 90 °C for 24 h under water saturated atmosphere. This process activates the geopolymer paste and hardens. Then, it was transferred to curing at 40 °C for 72 h under water-saturated atmosphere. Afterwards, geopolymer paste was cured at 40 °C for 72 h under open air. The processing condition was selected according to our previous studies [10,11]. The samples were coded as xRHRz where x stands for the amount of RHA used, z stands for the type of RHA used.

The universal testing machine (INSTRON 5566) was used to measure the compressive strength of samples at 7th day. The measurement was performed according to ASTM C39-04a. The average compressive strength was averaged from five samples. Bulk, apparent density, and porosity of the 7th day cured geopolymer was according to ASTM C642-06. Phase development was investigated using an X-ray Diffraction (XRD, PANanalytical, X'PertPRO). The XRD scans were operated at 5 to 70 °2θ, 0.0008 degree/step. The speed of XRD scans was set at 20 s/step.

Thermogravimetric analysis (MettlerToledo/851e) was performed to investigate mass changes of geopolymers upon heating. The samples approximately 20 – 25 mg was investigated in an alumina pan. The temperature range used was 20 – 1,200 °C and heating rate used was10 °C /min. Nitrogen gas was used in this experiment and used flow rate was 120 ml/minA subsection.

3. Results and discussion

The chemical compositions of FA, RHAR, RHA600 and RHA700 are shown in Table 1. The result showed that burning of RHA700 was relatively complete so only minor weight loss was observed. Phase analysis of the FA using an XRD technique in Fig. 1 showed that there existed 5 crystalline phases of anhydrite ($CaSO_4$), quartz (SiO_2), magnesioferrite (Fe_2MgO_4), hematite (Fe_2O_3), and lime (CaO). Particle size distribution result is for the starting materials are shown in Table 2.

Table 1. Chemical compositions of FA, RHAR, RHA600 and RHA700 as analyzed using an X-ray fluorescence. Noted that data for RHA600 and RHA700 was received from ref [8].

. Material	Chemicals Compositions (%weight)						
	SiO_2	Al_2O_3	Fe_2O_3	CaO	K_2O	SO_3	LOI
FA	29.765	14.248	19.074	24.545	2.309	9.226	0.281
RHAR	71.410	-	0.090	0.720	0.940	0.490	25.370
RHA600	89.850	-	0.060	0.340	1.930	0.260	7.410
RHA700	95.590	-	0.250	0.940	1.530	0.510	0.990

Table 2. Particle size distribution of FA and RHA

Material	**D10 (μm)**	**D50 (μm)**	**D90 (μm)**	**Mean (μm)**
FA	2.124	19.260	62.747	26.825
RHAR	6.070	40.156	115.227	58.727
RHA600	4.790	27.902	78.456	35.501
RHA700	7.212	40.595	95.223	46.414

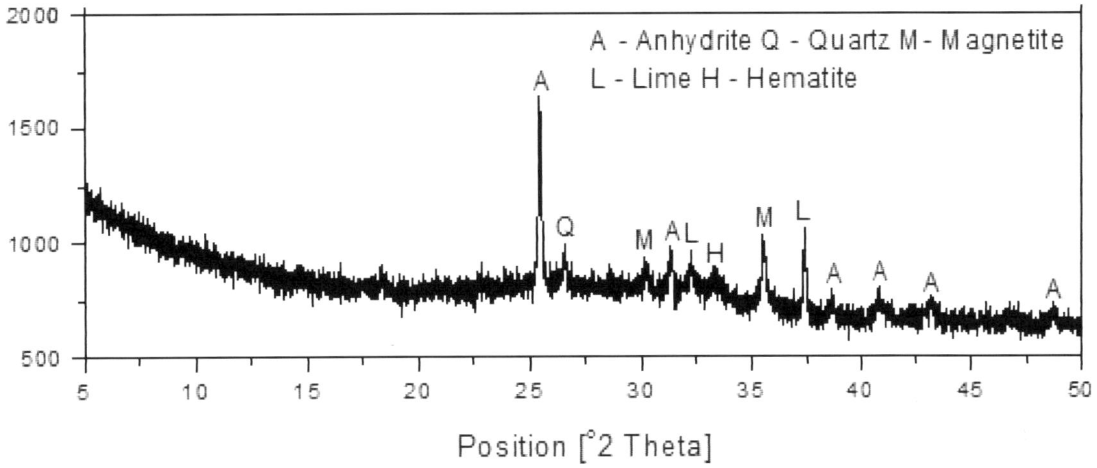

Figure 1. XRD spectrum of the FA used in this research

Thermograms (TGA) of the 3RHA700 composite is shown in Fig. 2. The analysis showed a large decrease of mass around 13% at 120 °C. Elimbi [9] reported that there were 3 types of water, namely adsorbed water, interstitial water, and hydroxyl water, existing in geopolymer. Elimination of these water occurred between 50-125 °C, 125-250 °C and 250 °C and 750 °C respectively. However, in our present result, only elimination of adsorbed water was mainly observed.

The XRD patterns of control and 3RHA700 are shown in Fig. 3 The broad hump observed approximately between 25 - 35° 2θ was due to the presence of the glassy phase. The dominant crystalline phases observed in control was quartz (SiO_2), calcium silicate hydrate ($CaSiO_3.xH_2O$), and lazurite ($Na_{8.56}(Al_6Si_6O_{24})(SO_4)_{1.56}\ Si_{0.44})$). The phases observed in the XRD spectrum for 3RHA700 were relatively similar to those of the control. It was noted that intensity of quartz increased while that of Lazurite decreased in the 3RHA700 as compared to the control. CHS was believed to transform from anhydrite in the presence of silicate reagents. The reduced quartz content was suggested its better dissolution in the presence of RHA.

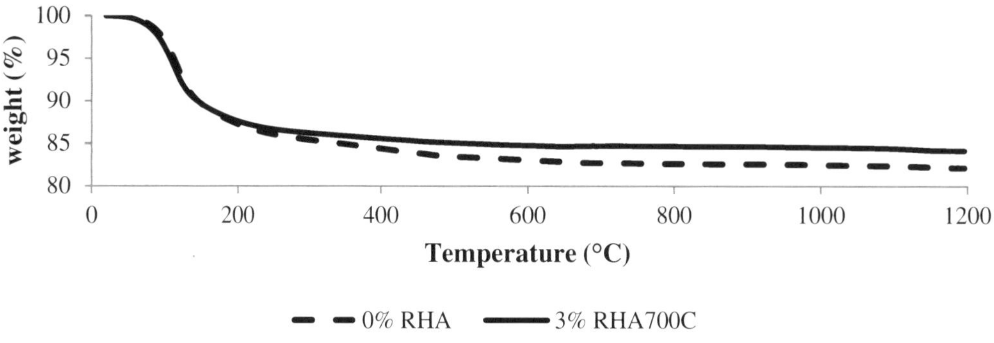

Figure 2. Thermogravimetric analysis for the control and 3RHA700 composite.

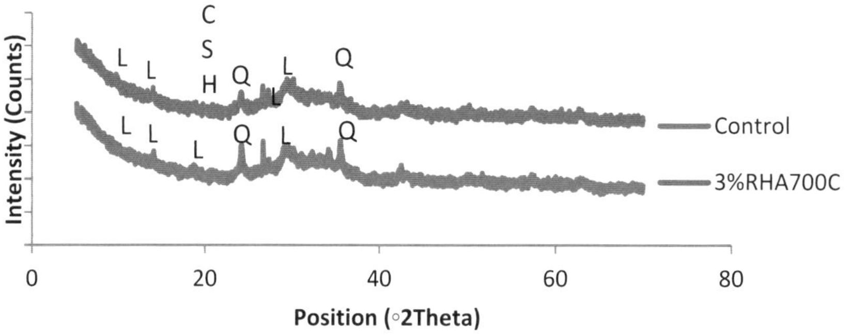

Figure 3. XRD patterns of the control and the 3RHA700 composite (CSH: Calcium Silicate Hydrate ($CaSiO_3.xH_2O$)), L: Lazurite ($Na_{8.56}(Al_6Si_6O_{24})(SO_4)_{1.56}S_{0.44}$)).

While bulk density did not show great variation among the sample types (~1.9 g/cm^3), apparent density was obviously different. The apparent density increased from 2.28 to around 2.36-2.41 g/cm^3 when 3% of RHA was incorporated regardless of the type of RHA. Further increasing RHA content to 5% resulted in a slight decrease of the apparent density and finally decreased to lower than the control in the 5% replacement.

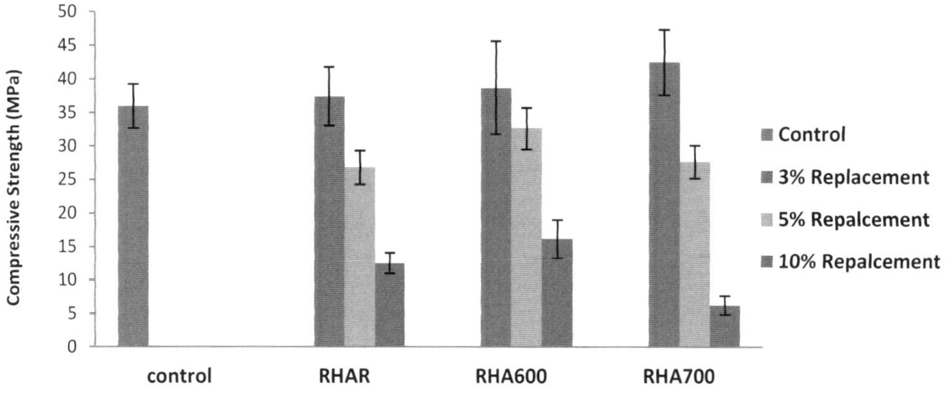

Figure 4. Compressive strength of FA-based geopolymers prepared with various RHA types and replacement contents.

The compressive strength for cured geopolymers was shown in Fig. 4. Compressive strength of the control was 35.91±3.30 MPa. Replacement FA with RHA in all types could obviously improve the compressive strength of the geopolymers with the increasing trend according to the SiO_2 content. The highest compressive strength of 42.52±4.86 MPa was found at 3% replacement with RHA700C. Increasing the replacement content damaged the compressive strength dramatically. Phoo-ngernkham reported that addition of SiO_2 could induce formation of calcium silicate hydrate (CSH) or calcium aluminosilicate hydrate (CAH) and sodium aluminosilicate hydrate (NAH) in high alkali and high calcium contents leading to enhanced strength [6]. Further increasing RHA contents resulted in low workability and difficulty in processing so compressive strength was reduced.

4. Conclusion

Preparation of geopolymer with replacement of FA with various types and contents of RHA was found to give interesting results. The results showed that the optimum content of RHA used should not be greater than 3% replacement. At this concentration, improvement of compressive strength with the SiO_2 content in RHA was observed. The enhanced compressive strength in the 3% replacement was suggested to be due to the increased apparent density and the better geopolymerisation of nanosilica in RHA. Further increasing RHA content resulted in a highly viscous mixture, workability and flowabilty became the key factors influencing the compressive strength.

Acknowledgement
Financial support from Mae Fah Luang University is highly appreciated.

References

[1] J. Davidovits, Global warming impact on the cement and aggregates industries, World Resour.Rev. *(*1994*)*, p. 263.
[2] J.C. Swanepoel, C.A. Strydom: Appl. Geochem. Vol. 17 *(8) (*2002*)*, p. 1143.
[3] T. Phoo-ngernkham, P. Chindaprasirt, V. Sata, S. Hanjitsuwan, and S. Hatanaka: Mater. & Design Vol. 55 (2014), p. 58.
[4] A. M. Al Bakria, H. Kamarudin, M. BinHussain, I. K. Nizar, Y. Zarina, and A. R. Rafiza: Physics Procedia Vol. 22(2011), p. 286.
[5] Y. Luna-Galiano, A. Cornejo, C. Leiva, L.F. Vilches, C. Fernandez-Pereira: Mater. De. Constr. Vol. 65*(*319*) (*2015*)*, p.e059.
[6] G. Görhan and G. Kürklü, Composites part b: Engineer. vol. 58(2014), p. 371,
[7] S. Naskar, A.K. Chakraborty: Perspect. Sci Vol. 8*(*2016*)*, p.273.
[8] D.Wattanasiriwech, N.Polpuak, P.Danthaisong, and S. Wattanasiriwech: J. Sci. Indus. Res. Vol. 67(6) (2008), p. 455.
[9] Aimbi, H. K. Tchakoute, M. Kondoh, and J. D. Manga: Ceram. Inter. Vol. 40(3) (2014), p. 4515.
[10] S. Wattanasiriwech, F. A. Nurgesang, D. Wattanasiriwech, P. Timakul: Ceram. Inter. Vol. 43(2017), p.16055.
[11] S. Wattanasiriwech, F. A. Nurgesang, D. Wattanasiriwech, P. Timakul: Ceram. Inter. Vol. 43(2017), p.16063.

Biodegradable fibers from poly (lactic acid)/poly (butylene succinate) blends

Kanokpun Panichsombat[1,a], Wisawachit Panbangpong[1,b], Nutthapong Poompiew[1,c], Pranut Potiyaraj[1,d]

[1]Department of Materials Science, Faculty of Science, Chulalongkorn University, Bangkok, 10330, Thailand

[a]p.kanokpun.tee@gmail.com; [b]wisawachitpan@gmail.com; [c]poompiew.n@chula.ac.th; [d]pranut.p@chula.ac.th

Abstract. PLA and PBS have more attention from many industries including textiles industry because there are biopolymer, which have eco-friendly. This research aims at fabricating biodegradable fibers from poly (lactic acid) (PLA) blended with poly (butylene succinate) (PBS). PLA and PBS were blended in a twin-screw extruder at various PBS contents 0 to 20 wt%. The blended fibers were then prepared by the melt spinning using a multifilament spinneret. The mechanical, thermal, and morphological properties were characterized by tensile testing, differential scanning calorimetry, and scanning electron microscopy. The results indicate that, at PBS content in PLA of 10 wt%, the obtained fiber has the optimal tensile properties. At this ratio, the blends also possesses the highest crystallinity among the blended fibers. Although these two biodegradable polymers are immiscible, but at the optimal ratio the PBS phase is well-dispersed in the PLA matrix phase.

1. Introduction

The polymer industry is growing steadily as polymers in various forms have been used as raw materials for fabrications of variety products. Despite its unique and customizable properties, the problems of plastic wastes and microplastics being accumulated in the environment due to its slow biodegradation of polymers raise a lot of concerns recently. One of their applications with high production capacities and consumer demands is for textile products which can be made from several types of polymers, especially synthetic fibers. Polyethylene terephthalate is the most consumed synthetic polymer which is used in textile products such as garments, home textiles, and automotive textiles. However, its non-biodegradability leads to a serious environmental problem. Consequently, biodegradable polymers get more and more attentions from textile industries. One of the most attractive biodegradable polymers is poly(lactic acid) (PLA) which is either chemically-synthesized from a biological materials or entirely bio-synthesized by living organisms [1]. PLA has been developed for food packaging applications as well as biomedical uses because it is non-toxic to human and also has superior mechanical properties and good thermal stability. However, PLA shows quite high glass transition temperature (T_g) that resulting in rigid and brittle fibers [2]. Thus, this study aims at reducing the rigidity and brittleness of PLA fiber via the polymer blending technique. Poly(butylene succinate) (PBS) was chosen to blend with PLA due to the fact that PBS possesses lower T_g than that of PLA thus illustrates more flexibility. PBS was blended with PLA at various ratios and successfully fabricated into multifilament fibers. The mechanical, thermal, and morphological properties of the blended fibers were then investigated. These properties should be tuned in order to expand the utilization of biodegradable polymers in textile applications.

2. Experimental

2.1. Materials
General purpose film grade PLA pellets were supplied by NatureWorks (Ingeo™ Biopolymer, 4043D) [3]. General purpose injection grade PBS pellets were supplied by Mitsubishi Chemical Performance Polymers (BioPBS™, FZ91PB) [4].

2.2. Preparation of PLA/PBS blend fibers
PLA and PBS pellets were dried at 60 ℃ for 24 hours. The pellets were compounded in a twin-screw extruder at PLA: PBS various ratios (100:0, 95:5, 90:10, 85:15, 80:20). The barrel temperatures were 150 – 160 ℃ from feed zone to metering zone with the screw speed of 32 rpm. The obtained blend pellets were kept drying at 60℃ for 24 hours. Fibers were prepared by the melt-spinning technique using a single screw extruder with barrel temperatures of 170 – 190 ℃ and the screw speed of 4 rpm. The molten polymers were extruded through a multifilament spinneret and were collected by a roller [1].

2.3. Characterizations
The characteristic functional groups of the samples were examined by Fourier transform infrared spectroscopy (FT-IR) with a Thermo scientific Nicolet 6700 (Tokyo, Japan). Polypyrrole coated zinc oxide nanoparticles were dried and coated with 10 nm of gold before being observed the surfaces using scanning electron microscopy (SEM: JSM-6480LV) at an acceleration voltage of 15 keV.

2.4. Testing methods
The mechanical properties of blended fibers are represented by the tensile testing where ASTM D3822 was followed using a Universal Testing Machine equipped with 5 N load cell at a crosshead speed of 5 mm/min. Samples were prepared as a single strand on C-shape paper which the length × width were 4.3 × 3.3 cm and the length of a single strand on C-shape paper was 2.3 cm. The thermal properties of blended fibers were tested by a differential scanning calorimeter (DSC) between 30 to 200 ℃ with the heating and cooling rate of 5 ℃/min under N_2 atmosphere. The fractured cross-sectional surface morphology of the blended extrudates were analyzed by a scanning electron microscope (SEM) with an acceleration voltage of 15 kV. The samples were prepared by breaking the extrudates after immersing in liquid nitrogen for 10 min.

3. Results and discussion
According to the tensile properties of neat PLA and PLA/PBS blends shown in Figure 1, the amount of PBS affects tensile properties significantly. The addition of PBS into PLA matrix actually reduces the maximum stress of the fibers but the increase in elongation at break is found higher than neat PLA and reaching the maximum when the percentage of PBS is 10 wt.%. Thus, the addition of PBS could reduce the brittleness of PLA [5]. However, at 5 wt.% of PBS, the elongation at break decreases significantly resulting from the incompatibility between PLA and PBS phases. When the amount of PBS increases beyond 10 wt%, the elongation at break decreases probably because of PBS agglomeration.

IOP Conf. Series: Materials Science and Engineering **600** (2019) 012004 doi:10.1088/1757-899X/600/1/012004

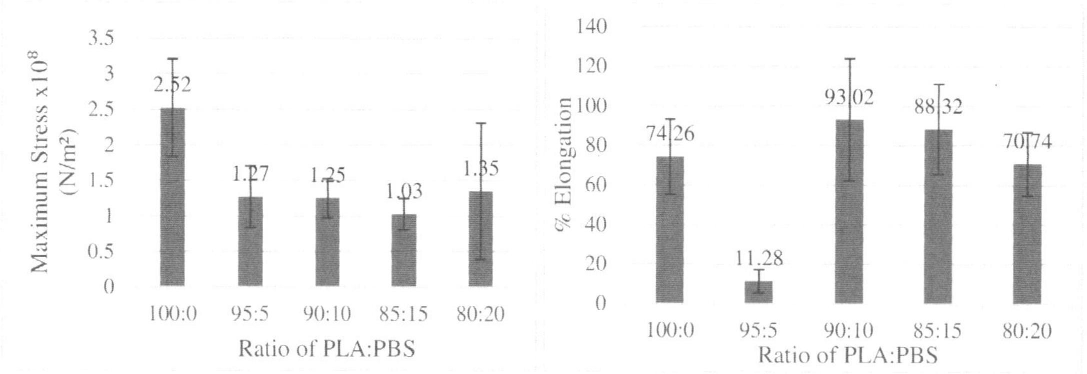

Figure 1. (a) maximum stress (N/m²) and (b) elongations percentage of various PLA:PBS

To study the influence of incorporating PBS into PLA on morphology, DSC was employed to measure the crystallinity difference between neat PLA and PLA/PBS blends. The T_g for neat PLA and PLA/PBS blends does not change significantly while the melting temperature (T_m) slightly shift lower when PBS content increase (Figure 2.).From Table 1, the crystallinity percentage of PLA/PBS blends is reduced comparing with neat PLA because of the interruption of PLA chain rearrangement. However, at 10 wt% PBS showed more crystallinity than other bended fibers. This affects the tensile properties as explained earlier.

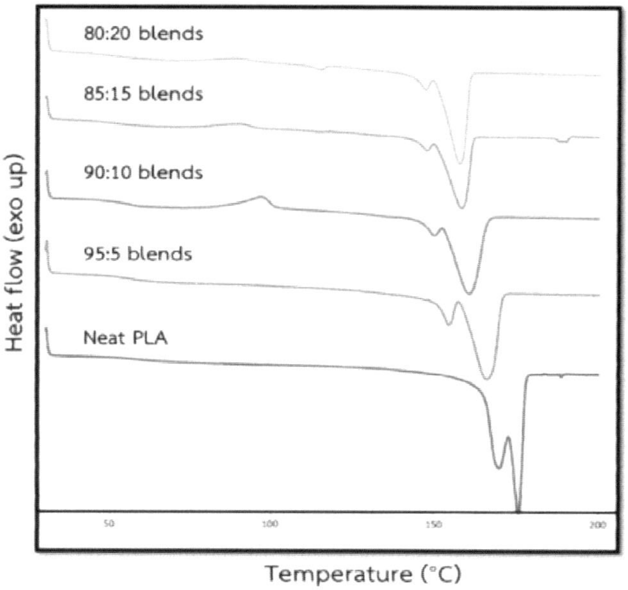

Figure 2. DSC second heating thermographs for neat PLA and PLA/PBS blends

Table 1. Thermal properties of neat PLA and PLA/PBS blends

Sample PLA:PBS	T_g (°C)	T_c (°C)	T_m (°C)	X_c (%)
100:0	63.53	115.32	175.69	33.26
95:5	62.38	106.39	166.08	12.42
90:10	58.88	100.38	160.41	18.19
85:15	60.54	97.53	158.18	10.19
80:20	58.14	97.54	157.66	10.24

The compatibility of PBS in the PLA matrix was characterized by scanning electron microscopy (SEM). The results indicate that PBS phase obviously separates from PLA phase because the incompatibility of PLA and PBS(Figure 3.). At 5 wt.% of PBS, poor distribution and agglomeration of PBS are observed while at 10 wt.% of PBS, PBS phase has regularly size and is well distributed in PLA phase. When the amount of PBS increases, PBS phase becomes larger and agglomeration is pertinent thus blocking PLA chain to rearrange causing the low crystallinity percentage and lower mechanical properties. In figure 4, the fibers surface of 5 wt.% of PBS are rough more than others because this ratio have high incompatibility. The high incompatibility and poor distribution are cause of very low mechanical properties of fibers.

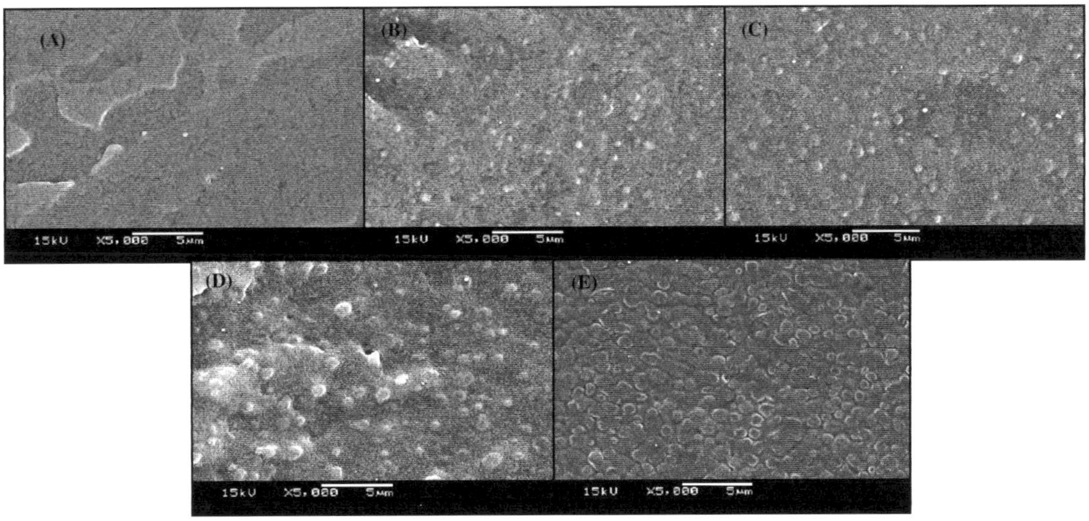

Figure 3. SEM micrographs of the cross section of PLA/PBS blends (A) neat.PLA (B) 95:5 blends (C) 90:10 blends (D) 85:15 blends and (E) 80:20 blends at 5000x and scale bar is 5 μm

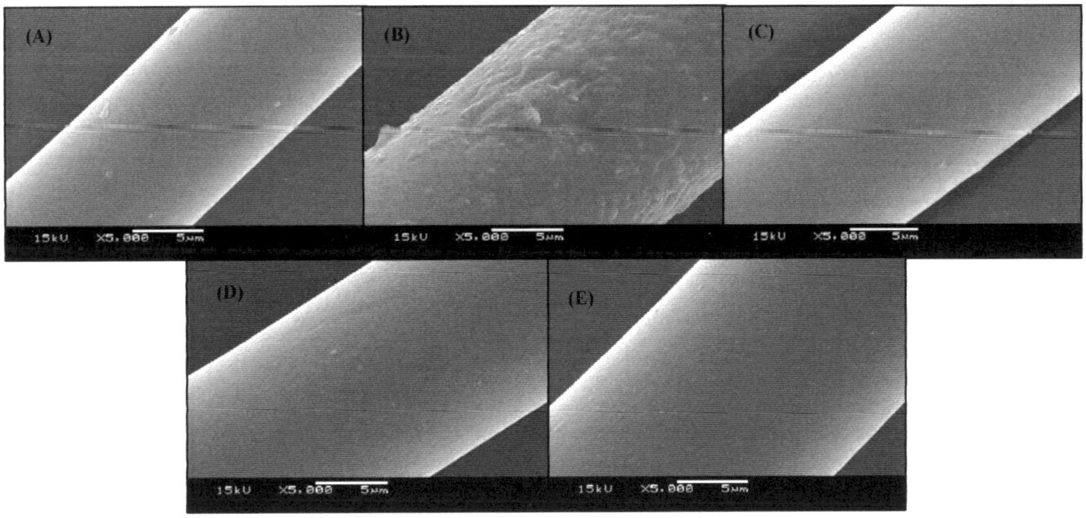

Figure 4. SEM micrographs of the surface of PLA/PBS fibers (A) neat.PLA (B) 95:5 blends (C) 90:10 blends (D) 85:15 blends and (E) 80:20 blends at 5000x and scale bar is 5 μm

IOP Conf. Series: Materials Science and Engineering **600** (2019) 012004 doi:10.1088/1757-899X/600/1/012004

4. Conclusion

The mechanical, thermal, and morphological properties of PLA/PBS blended fibers were thoroughly investigated. It can be concluded that PLA/PBS blended fiber at the ratio of 90:10 gives the most suitable properties for textile fiber application. While the blends have lower maximum stress than neat PLA, the elongation at break increases at approximately 20% when the amount of PBS in PLA is 10 wt.%. The DSC measurement indicates that the presence of PBS does have a remarkable impact on the crystallinity of the PLA/PBS blends fibers. In addition, the SEM confirms that, when the amount PBS is 10 wt.%, the PBS phase is well-distributed giving a high crystallinity percentage as well as good mechanical properties suitable for the textile fiber application. In the future, we would to develop mechanical and thermal properties to the blends with graphene particles [6].

References

[1] Jompang L, Thumsorn S, Wong-On J, Surin P, Apawet C, Chaichalermwong T, Kaabbuathong N, O-Charoen N and Srisawat N. Poly(lactic acid) and Poly(butylene succinate) Blend Fibers Prepared by Melt Spinning Technique. *Energy Procedia*. 2013; 34: 493-499

[2] Rasal RM, Janorkar AV, Hirt DE. Poly(lactic acid) modifications. *Progress in Polymer Science*. 2010; 35: 338-356

[3] Ingeo™ Biopolymer on http://www.natureworksllc.com

[4] BioPBS™ on http://www.pttmcc.com

[5] Valapa R, Hussain S, Iyer PK, Pugazhenthi G and Katiyar V.Influence of graphene on thermal degradation and crystallization kinetics behavior of poly(lactic acid). *Journal of Polymer Research. 2015; :22 189-175*

[6] Wang X, Yang H, Song L, Hu Y, Xing W and Lu H. Morphology, mechanical and thermal properties of graphene reinforced poly(butylene succinate) nanocomposites. *Composites Science and Technology. 2011; :72 6-1*

Preparation of bismuth ferrite as photo-supercapacitive electrode

H Haromae[1,3] and P Pattananuwat[1, 2, 3, *]

[1]Department of Materials Science, Faculty of Science, Chulalongkorn University, Bangkok, 10330, Thailand
[2]Research Unit of Advanced Materials for Energy Storage, Chulalongkorn University, Bangkok, 10330, Thailand
[3]Center of Excellence on Petrochemical and Materials Technology, Chulalongkorn University, Bangkok, 10330, Thailand

E-mail: *prasit.pat@chula.ac.th

Abstract. In this work, we report a novel concept of photo-assisted charging supercapacitive electrode of bismuth ferrite (BFO). The thin film layer of bismuth ferrite electrode is synthesized by electrodeposition technique. The effect of bismuth (III) nitrate/iron (III) nitrate ratio on the capacitive behavior of BFO electrode is investigated. FE-SEM demonstrates the morphology development of flake nanostructure of BFO at the equivalent ratio. The photo-supercapacitive electrode can deliver the highest specific capacitance of 220.50 F g^{-1} at 10 mV sec^{-1} under LED light irradiation which is higher than that obtained from dark conditions. This photo active layer can integrate function with the capacitive layer inducing photo-charging mechanism, thus it has great potential of this material electrode for photo-assisted charging supercapacitors application.

1. Introduction

Nowadays, there have been rapid developments in portable electronic devices and hybrid electric vehicles. Supercapacitors (SCs) are becoming the most attention as power energy supplies because of their high-power energy density, long life cycle, and fast charge/discharge system. Much effort has been devoted the materials for high-performance supercapacitors including transition metal oxides. Furthermore, the challenge to develop supercapacitors is not only to increase their electrochemical performance but also to elevate the function-level of supercapacitors by combining with others energy technology. Recently, the new concept is purposed as the photo-assisted charging supercapacitor devices, which directly integrate the system of harvesting and collecting energy in one electrode. Only a few reports attempted to develop electrode by using the photoactive/redox materials. Some materials classes of metal oxide such as $Co(OH)_2$ is purposed as photo-assisted charging supercapacitors. Its capacitive properties can gain the more charge transfer under blue light irradiation, but the capacitive efficiency is still not desire to achieve high power energy [1]. Bismuth ferrite (BFO) is one of promising semiconductor materials for supercapacitors with rhombohedral distorted perovskite structure. It has a narrow band gap between 2.3 and 2.8 eV with high chemical stability [2]. Theoretically, the synthesized nanoflake bismuth ferrite thin film via electrodeposition technique can yield the specific capacitance of 72.2 F g^{-1} at a current density of 1 A g^{-1} [3]. Moreover, bismuth ferrite can be improved photocurrent response in photovoltaic applications [4]. Interestingly, this material classes reveal the multipotential efficiency for inducing charge redox reaction under the appropriate light energy. Also, the study of its photo-sensing inducing capacitive properties still has not been widely addressed. Thus, in the present study, the photoactive energy storage materials using bismuth ferrite (BFO) is purposed. The effect of bismuth (III) nitrate to iron (III) nitrate ratio on photo-electrocapacitive property is studied. Photo-assisted charging supercapacitors of BFO electrode is evaluated by cyclic voltammetry (CV) and galvanostatic charge-discharge (GCD) under dark conditions and LED light irradiation.

2. Experimental section

2.1. Preparation of bismuth ferrite electrode
Bismuth ferrite thin films were electrodeposited on FTO substrate by applying voltage - 1.5 V for 900 second from the mixture aqueous electrolyte solution of 0.1 M bismuth *(*III*)* nitrate and 0.1 M Ferrite *(*III*)* nitrate at various ratio of 1:1, 1:2 and 2:1. The adding of 0.1 M tartaric acid solution was used to form complex substance and the aqueous NaOH is used to adjust pH. The various ratios of bismuth (III) nitrate to iron (III) nitrate at 1:1, 2:1 and 1:2 were named as BFO11, BFO21, BFO12, respectively.

2.2. Characterizations
The surface morphology of the samples was observed by field emission scanning electron microscope *(*FE-SEM, HITACHI SU-8010*)* at an acceleration voltage of 10 keV. The optical characteristics was examined using UV-Vis Spectrophotometer *(*LabTech BlueStar B*)*.

2.3. Electrochemical measurements
To evaluate the electrochemical characteristics, cyclic voltammetry and galvanostatic charge/discharge were applied to conventional three-electrode configuration. The as-prepared samples, Ag/AgCl and platinum plate were used as working electrode, counter electrode and reference electrode, respectively. For cyclic voltammograms, the response current densities of the as-prepared electrode samples were recorded in 1.0 M KOH between -1.0 to -0.4 V *(*versus Ag/AgCl*)* from 10 to 100 mV/sec. The galvanostatic charge-discharge *(*GCD*)* of the as-prepared electrode samples were also recorded in 1.0 M KOH at current density range of 1 to 10 A g^{-1} from -1.0 to -0.4 V with open circuit potential using PGSTAT 12 from 50 mHz to 10 KHz with alternate current amplitude of 5 mV. All the samples test was tested under dark conditions and LED light irradiation.

3. Results and discussion
The surface morphology of the as-prepared BFO electrode are investigated by field emission scanning electron microscopy *(*FE-SEM*)*, as shown in fig. 1*(a)* - *(c)*. Obviously, BFO surface demonstrates the development of flake structures depending on the ratio of Bi $(NO_3)_3$ to Fe $(NO_3)_3$. At Bi $(NO_3)_3$/ Fe $(NO_3)_3$ ratio of 1 to 2, BFO reveals the typical structure of the dense flake morphology as seen in fig. 1*(a)*. The increase the ratio of Bi $(NO_3)_3$ to Fe $(NO_3)_3$ up to 1 to 1 affects to increase the dense growth of anisotropic flake nanostructure surface as seen in fig.1*(*b*)*. The presence of an aggregate irregular flake structures is observed as seen fig. 1*(c)* resulting from the increase Bi $(NO_3)_3$/Fe $(NO_3)_3$ ratio at 2 to 1, implying the trend of the occurring of the aggregate irregular shape with high content of Bi $(NO_3)_3$. Additionality, the energy dispersive X-ray spectroscopy *(*EDX*)* signal pattern of BFO11 electrode deposited onto FTO substrate is used to confirm the element surface of BFO11, consisting of Bi, Fe and O elements as displayed in fig. 1d. This presence of element ratio of Bi, Fe and O is identical close to the theoretical value of BFO. The optical properties of BFO11, BFO21 and BFO12 are characterized by UV-visible spectrophotometer as shown in Fig. 2. All BFO samples show the absorption spectra in the ultraviolet range, responding strongly in the range between 280 to 400 nm

and a sharp decline at wavelengths over 400 nm. Obviously, BFO11 shows the strongest optical absorption, expecting to possess the highest photocurrent response in photovoltaic applications.

Figure 1. FE-SEM images of *(a)* BFO12, *(b)* BFO11, *(c)* BFO21 and *(d)* EDX spectra of BFO11

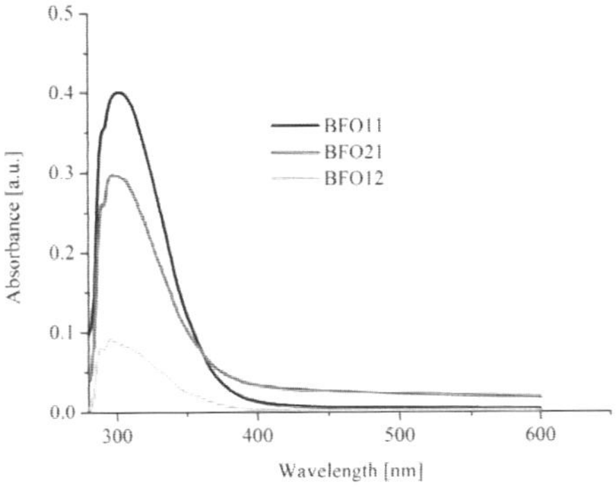

Figure 2. UV-Visible absorption spectra of BFO11, BFO21 and BFO12 samples

The electrochemical performance of BFO sample electrodes are evaluated using CV and GCD under dark condition and LED light irradiation. As seen in Fig. 3a and b, the electrodes of BFO11 displays the largest CVs area under both conditions than that obtained from BFO21 and BFO12 electrodes, indicating the highest activity. All GCD curves in fig. 3 c and d reveal an initial drop in the voltage during discharging owning to the internal resistances of low conductivity surface of BFO. The typical battery-like features and good pseudocapacitive behavior can be easily observed during discharge process. GCD results for all BFO electrodes are similar tendency with CV. BFO11 exhibits the longest discharge time under both conditions. Thus, it can be concluded that the optimum ratio of Bi $(NO_3)_3$/Fe $(NO_3)_3$ for the highest electrochemical performance under both dark and LED light condition is at ratio of 1 to 1. These results also correspond agreement with FE-SEM results supporting that the high active surface of flake structure of BFO11 can provide a facile shortcut path for charge transfer increasing a more efficient electrocapacitive properties [5]. The calculated specific capacitance values from GCD of the all as-prepared BFO electrodes under both conditions are showed as in fig 3e. BFO11 electrode under LED light can provide the more efficiency of 11% higher than that obtain from dark condition (168.50 to 152.83 F g^{-1}). Interestingly, the increment of capacitance and current under LED light irradiation can occur from both the photo-charging mechanism and redox reaction. BFO is excited with LED light and then changed into the hole (h^+) and electron pair (e^-), thereby providing the more charge under LED light irradiation [6].

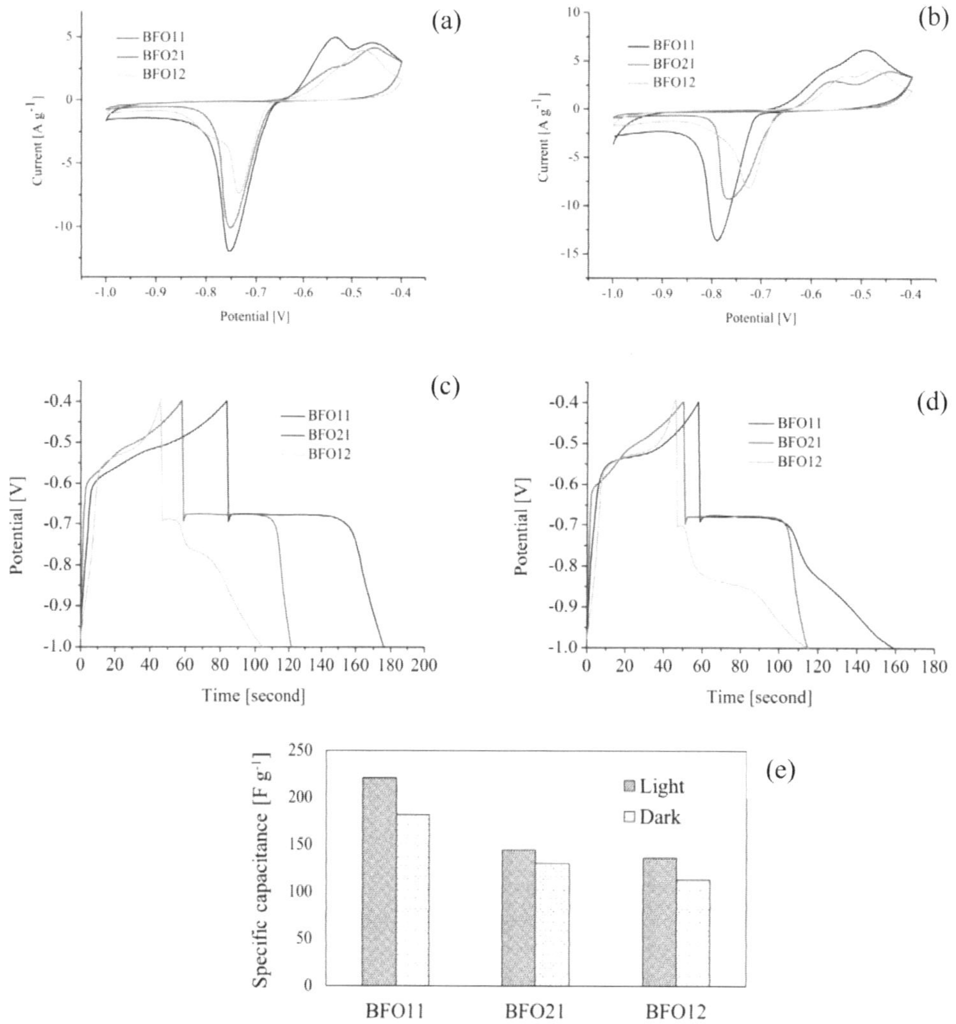

Figure 3. CV curves of various samples at 10 mV sec^{-1} in 1.0 M KOH (a) under dark condition (b) under LED light irradiation; GCD curves of various samples at 1 A g^{-1} in 1.0 M KOH (c) under dark condition (d) under LED light irradiation (e) specific capacitance of various samples at 10 mV sec^{-1}

The electrochemical performance of BFO11 electrode under dark and LED light irradiation at various scan rate is also evaluated as shown in fig. 4a and b. The good rate capability of BFO11 electrode with capacitance preservations around 31.07% (182.17 to 56.6 F g^{-1}) for dark condition and 32.27% (220.50 to 71.15 F g^{-1}) for LED light irradiation as the increasing of scan rate from 10 to 100 mV sec^{-1}, confirming that the better performance retention of BFO11 under LED light irradiation. GCD curve results of BFO11 under dark condition and LED light irradiation at various current densities also show the similar tendency results. The specific capacitance of BFO11 is 152.83 F g^{-1} and 168.50 F g^{-1} at 1 A g^{-1} for dark condition and LED light irradiation, respectively. It can maintain the capacitance preservations around 3.49% and 2.57% as current density from 1 to 10 A g^{-1} under dark condition and the LED light irradiation, respectively.

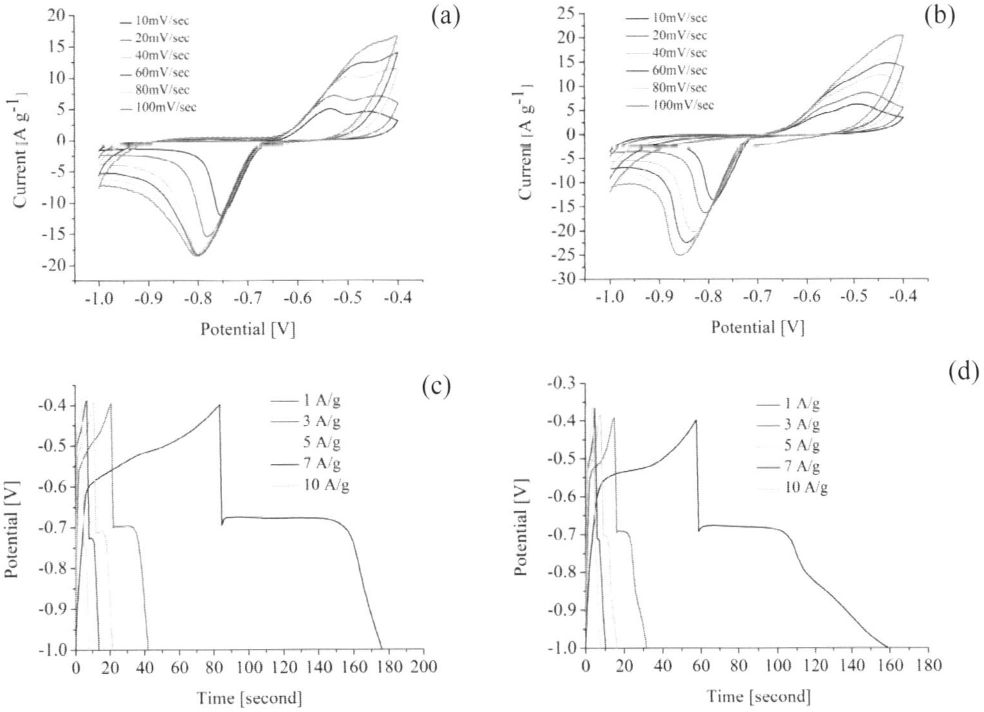

Figure 4. CV curves of BFO11 at different scan rates in 1.0 M KOH (a) under dark condition (b) under LED light irradiation; GCD curves of BFO11 at different current densities in 1.0 M KOH (c) under dark condition (d) under LED light irradiation

4. Conclusions

In summary, the as-electrodeposited bismuth ferrite with an equal ratio of bismuth *(III)* nitrate and iron *(III)* nitrate can provide the highest specific capacitance values under LED light irradiation due to the synergistic effects of pseudocapacitor mechanism and photo-charge generation mechanism of bismuth ferrite materials. This novel class material is purposed as charging photoactive supercapacitor and developed as promising materials for the next generation of photo-supercapacitor energy technology.

References

[1] S. Kalasina, P. Pattanasattayavong, M. Suksomboon, N. Phattharasupakun, J. Wutthiprom and M. Sawangphruk: Chem. Commun. Vol. 53(4) (2017), p. 709-712

[2] D. Tiwari, D.J. Fermin, T.K. Chaudhuri and A. Ray: J. Phys. Chem C. Vol. 119 (2015), p. 5872-5877

[3] V.V. Jadhav, M.K. Zate, S. Liu, M. Naushad, R.S. Mane, K.N. Hui and S.H. Han: Appl. Nanosci. Vol. 6(4) (2015), p. 511-519

[4] R. Agarwal, Y. Sharma, F. Mendoza, G. Morell and R.S. Katiyar: ECS Transactions. Vol. 75 (31) (2017), p. 1-9

[5] D. Moitra, C. Anand, B.K. Ghosh, M. Chandel and N.N. Ghosh: ACS Appl. Energy Mater. Vol. 1(2) (2018), p. 464-474

[6] T. Gao, Z. Chen, Y. Zhu, F. Niu, Q. Huang, L. Qin, X. Sun and Y. Huang: Mater Res Bull. Vol. 59 (2014), p. 6-12

IOP Conf. Series: Materials Science and Engineering **600** (2019) 012006 doi:10.1088/1757-899X/600/1/012006

Ferrocene-Graphene nanocomposite modified glassy carbon electrode for amaranth detection

Chutimon Akkapinyo[1, a], Dujduan Waraho-Zhmayev[1, b] and Rungtiva P. Poo-arporn[1, c]

Biological Engineering Program, Faculty of Engineering, King Mongkut's University of Technology Thonburi, Bangkok 10140, Thailand

[a]chutimon.akk@gmail.com; [b]dujduan.war@mail.kmutt.ac.th; [c]rungtiva.pal@kmutt.ac.th

Abstract. A non-covalent nanohybrid of Ferrocene (Fc) with grapheme (Gr) via π-π interaction was proposed for amaranth detection. The synthesized Fc-Gr composite was modified on glassy carbon electrode (GCE) and characterized by cyclic voltammetry (CV) as well as differential pulse voltammetry (DPV). Cyclic voltammogram and differential pulse voltammogram demonstrated successful fabrication of Fc-Gr composite-modified GCE for amaranth detection. DPV was used to obtain a linear range of amaranth detection. The linear range of amaranth detection with a modified sensor was expressed from 10 to 250 μM of amaranth. The limit of detection was obtained at 155 nM of amaranth. Based on these results, the developed sensor provides a good alternative for the determination of amaranth. Moreover, the developed sensor can hopefully be utilized and adapted for detection of other substances.

Keywords: Amaranth; Ferrocene; Graphene; Cyclic voltammetry; Differential pulse voltammetry

1. Introduction

In recent years, several food additives have been permitted in the food industry. Amaranth is one of them. Amaranth is a synthetic food colouring substance that provides red colorant. It has been widely used to make more fascinating food. However, excessive consumption of amaranth has become a concern because it has the potential to cause many adverse health effects such as nausea, vomiting, dizziness. Further, it is hazardous to internal organs such as the liver and kidneys. Moreover, amaranth has been reported to be hazardous to fetuses [1]. Therefore, the European Food Safety Authority (EFSA) defined an acceptable daily intake of amaranth at 0.15 mg/kg of body weight per day [2]. As amaranth's structure contains azo group -(N=N)- which can be reduced at the electrode surface, many researches have proposed electrochemical methods with different modification electrodes to determine amaranth levels in food [3-7].

Graphene is comprised of a single layer of carbon atoms arranged in a honeycomb structure. Due to its properties including excellent conductivity, electron transfer promotion and large surface area [8, 9],

it has been widely used to modify electrodes in the electrochemistry field to enhance sensor effectiveness. However, it is prone to restack due to strong π-π interaction between graphene sheets [10, 11]. Therefore, it is imperative to decorate graphene to enhance its dispersibility and stability.
Ferrocene (Fc) consists of two cyclopentadienyl rings bound on opposite sides of a central iron atom[8]. A π-electron system also exists in the molecule [12]. Thus, it is possible to intercalate Fc between layers of graphene sheet via π-π interaction. There have been numerous reports concerning the non-covalent hybridization of Fc on graphene oxide [10, 13, 14], reduced graphene oxide [8, 9, 15] and carbon nanotubes [16] to enhance electrochemical catalytic properties. However, few reports exist concerning research on the hybridization of Fc on graphene [11].
In this work, facile synthesis of Fc functionalized graphene is proposed. Fc-Gr nanocomposite was modified on glassy carbon electrode for amaranth detection. The modified electrode was characterized by cyclic voltammetry (CV) and differential pulse voltammetry (DPV).

2. Experiment details

2.1. Reagents and apparatus
Amaranth and Ferrocene (Fc) were purchased from Sigma Aldrich. Pristine graphene with diameter between 20-100 nm was obtained from cheaptubes.com. Electrochemical measurement was performed on an AutoLab PGSTAT128N system (EcoChemie, Netherlands). The three electrodes system consisted of modified glassy carbon electrode; GCE (3 mm in diameter) as working electrode, Ag/AgCl as reference electrode and Platinum wire as counter electrode. 0.1M phosphate buffer solution (PBS) pH 7.0 containing 0.1M KCl was used as electrolyte for electrochemical measurement. All solutions were prepared with deionized (DI) water and purged with nitrogen gas for 30 minutes to eliminate dissolved oxygen.

2.2. Synthesis of ferrocene-graphene (Fc-Gr) nanocomposites
20 mg of graphene was dispersed in 10 mL of ethanol by sonication for 30 min. 80 mg of Fc was dissolved in 10 mL of ethanol. Then, the dissolved Fc solution was added into a dispersed graphene solution and sonicated for 4 h. After that, the Fc-Gr nanocomposites were collected by centrifugation at 13,000 rpm for 10 min. The pellet was washed several times with ethanol to eliminate non-hybridized Fc. Finally, the precipitated Fc-Gr nanocomposites were dried at 37°C and stored under dry conditions.

2.3. Preparation of ferrocene-graphene (Fc-Gr) modified glassy carbon electrode
The bare GCE was polished on 0.05 alumina powder until a mirror-like surface was obtained. Then, it was rinsed with DI water followed by sonication in 70% ethanol and DI water for 5 min. The synthesized Fc-Gr nanocomposites were dispersed in DI water to a final concentration of 1 mg/mL. Fc-Gr modified GCE was prepared by dropping 5 μL of Fc-Gr solution on the surface of cleaned GCE followed by drying at room temperature for at least 1 h.

3. Results and discussion

3.1. Characterization of electrode modification
To characterize the assembly process, CV was studied on bare GCE, Gr modified electrode and Fc-Gr modified GCE. Fig.1 represents cyclic voltammograms of three types of electrode in 10 mM $Fe(CN)_6^{3-/4-}$ containing 0.1 M KCl at a scan rate of 50 mV/s. The presented current was the signal of $Fe(CN)_6^{3-/4-}$ in electrolyze on different electrodes. The current increased from bare GCE, Gr modified electrode and Fc-Gr modified GCE. Increasing current was the effect of Gr on electrode. The large surface area can help to increase the electron transfer rate and conductivity performance of electrode, which affects the current increment. At the same time, intercalation of Fc molecule can prevent restacking of graphene and help to form a larger conjugated system. Therefore, it can accelerate the electron transfer

rate on Fc-Gr modified GCE. The highest current on Fc-Gr modified GCE demonstrates that Fc is hybridized on Gr sheet by π-π interaction and the composite is stacked on GCE surface by physical absorption.

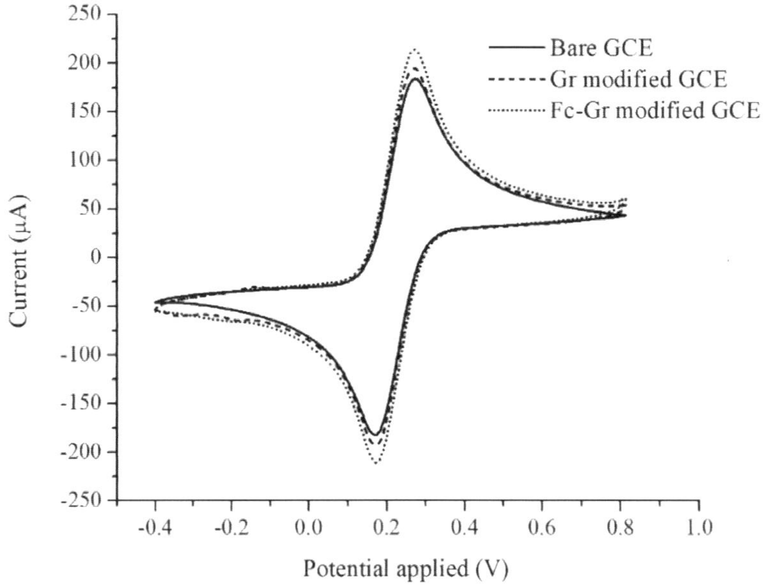

Figure 1. Comparison of cyclic voltammograms for bare GCE (solid line), Gr-modified GCE (dash line) and Fc-Gr modified GCE (dot line) in 10 mM $Fe(CN)_6^{3/-4-}$ containing 0.1 M KCl at a scan rate of 50 mV/s.

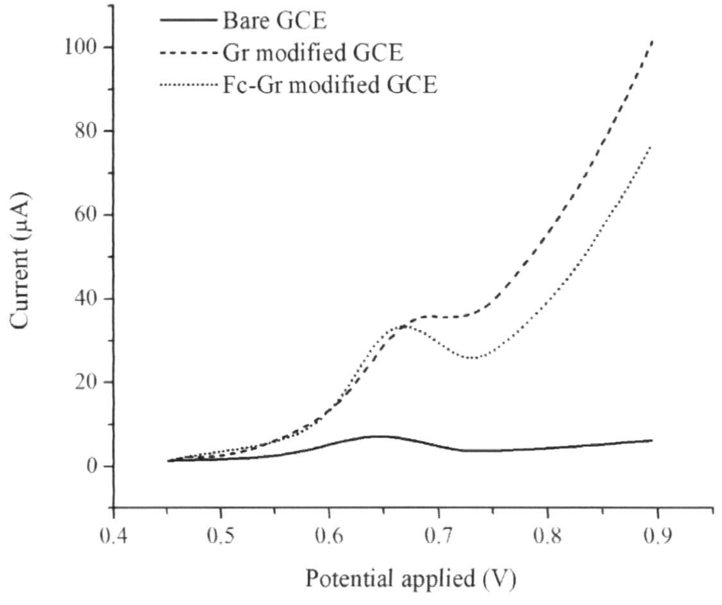

Figure 2. Comparison of differential pulse voltammograms for 10 μM amaranth on bare GCE (solid line), Gr modified GCE (dash line) and Fc-Gr modified GCE (dot line) in 0.1 M PBS containing 0.1M KCl.

3.2. Electrochemical behaviour of amaranth on modified electrode

The electrochemical behaviour of amaranth was tested on three different electrodes to study the performance of the modified electrodes. Differential pulse voltammograms of 10 µM amaranth in 0.1 M PBS containing 0.1M KCl were recorded on bare GCE, Gr modified GCE and Fc-Gr modified GCE at a scan rate of 50 mV/s. Fig. 2 demonstrates that the anodic peak current of amaranth was observed on all three types of electrode with slightly different potential. Anodic peak current of amaranth appeared at potential 0.64, 0.67 and 0.66 on bare GCE, Gr modified GCE and Fc-Gr modified GCE, respectively. The results were consistent with the characterization of electrode modification by CV. The highest current was obtained from Fc-Gr modified GCE, while the current of Gr modified GCE was higher than bare GCE. The higher current was the result of Gr and Fc-Gr on the electrode. The increment of the anodic peak current can prove that Fc-Gr composite helps to accelerate electron transfer between amaranth and electrode.

3.3. Determination of linear range and limit of detection

Standard solutions of amaranth with various concentrations at 0.41, 0.82, 1.23, 1.64, 2.05, 2.46, 2.87, 3.28, 3.69, 4.10, 6.15, 8.20, 10.25, 12.30 and 14.35 mM were prepared in 0.1 M PBS containing 0.1 M KCl. 250 µL of each solution was injected into an electrochemical cell to obtain the desired concentration in bulk for analysis. DPV of each concentration was studied at potential 0.4 to 0.9 V and at step potential 0.01 V. The current signal increased when amaranth concentration was increased. However, current signal tended to be stable when amaranth concentration reached 300 µM as demonstrated in Fig.3. Thus, the linear range was obtained at amaranth concentration ranging from 10 to 250 µM. The linear equation can be expressed as $y = 0.818x - 4.327$ ($R^2 = 0.971$). The detection limit was calculated from the ratio between 3 times the standard deviation of the blank (n = 10) and slope from the linear range. A detection limit of 155 nM was obtained from this developed sensor.

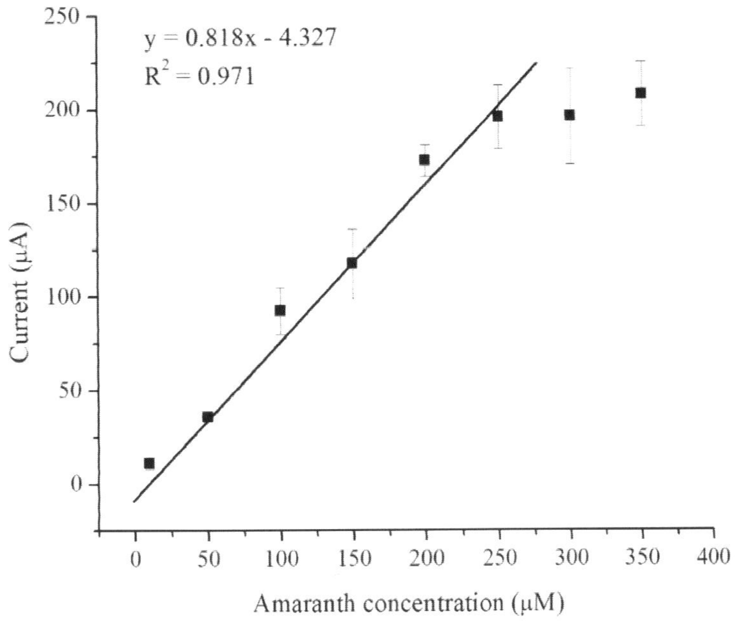

Figure 3. Linear relationship between the obtained current and concentration of amaranth in a range from 10 to 250 µM (n=3).

However, when comparing the modified sensor with a previous work that applied Fc-Gr modified GCE for chloromycetin detection [11], our modified sensor still has an advantage of facile synthesis.

Although, both our modified and the chloromycetin sensors utilized the π-π interaction to intercalate between graphene layer but their synthesized composite was not immobilized onto GCE via physical absorption. It was mixed with graphite and polyvinylidene fluoride (PVDF) for the formation of conductive polymer to coat on GCE. Although, PVDF has good binding property but it has poor conductivity. Thus, excessive PVDF can inhibit electron transfer on the electrode. Moreover, the modified sensor can be optimized by other factors, for examples, pH of electrolyte, scan rate or accumulation time. These factors influence the catalytic reaction of amaranth, therefore the developed sensor can be optimized to improve the performance of the sensor.

4. Conclusion

In summary, Fc-Gr composite modified GCE was successfully used as a sensor for amaranth detection. The synthesis of the composite was easily successful by π-π interaction between graphene and Fc without hazardous reagent. The preparation and fabrication of the sensor can be achieved with an easy method that is cost-effective and environmental friendly. The developed sensor worked well and demonstrated a wide linear range with low detection limit. However, this work did not test using a real sample. Thus, there are many other factors that should be optimized and adjusted by future research in order to obtain greater performance from the sensor.

Acknowledgment

This work was supported by the research subsidy fund of fiscal year 2019 from King Mongkut's University of Technology Thonburi.

References

[1] Chandran, S., Lonappan, L.A., Thomas, D., Jos, T., and Kumar, K.G) .2014(, *Food Analytical Methods*, Vol .7, p .741-746.
[2] ESFA Panel on Food Additives and Nutrient Sources added to Food)ANS) .(2010(, *EFSA Journal* .
[3] Jing, S., Zheng, H., Zhao, L., Qu, L., and Yu, L) .2017(, *Food Analytical Methods*, Vol .10, p . 3149–3157.
[4] Q .Hu, Q., Gao, H., M .Wang, Y., Ma, W., and M .Sun, D) .2018(, *Journal of Analytical Chemistry*, Vol.73, p.817-823.
[5] He, J-.L., Kou, W., Li, C., Cai, J-.J., Kong, F-.Y., and Wang, W) .2015(, *International Journal of Electrochemical Science,* Vol .10, p .10074 -10082.
[6] Wang, P., Hu, X., Cheng, Q., Zhao, X., Fu, X., and Wu, K) .2010(, *Agriculture and Food Chemistry*, Vol .58, p .12112-12116.
[7] Huang, W., Zhang, M., and Hu, W) .2017(, *Ionics*, Vol .23, p .241–246.
[8] Karthick, A., Thangappan, R., Arivanandhan, M., Gnanamani, A., and Jayavel, R).2017(, *Journal of Inorganic and Organometallic Polymers and Materials*, Vol .28, p .1021–1028.
[9] Rajesh, K., Kumar, D.R., and Santhanalakshmi, J., *International Journal of Innovative Research in Science & Engineering*.
[10] Li, C., Qiu, X., Deng, k., and Hou, Z) .2014(, *Analytical Methods*, Vol .6, pp .9078-9084.
[11] Hu, L., 2018, *Int .J .Electrochem .Sci*, Vol .13, p .396 –409.
[12] Gao, Y., Hu, G., Zhang, W., Ma, D., and Bao, X) .2011(, *Dalton Transactions*, Vol .40, p . 4542-4547.
[13] Yang, K., Qi, L., Gao, Z., Zu, X., and Chen, M) .2014(, *Analytical Letters*, Vol .47, p .2266-2280.
[14] Ai, W., Du, Z-.Z., Liu, J-.Q., Zhao, F., Yi, M-.D., Xie, L-.H., Shi, N-.E., Ma, Y-.W., Qian, Y., Fan, Q-.L., Yu, T., and Huang, W) .2012(, *RSC Advances*, Vol .2, p .12204-12209.
[15] Deng, K., Zhou, J., and Li, X) .2013(, *Electrochimica Acta*, Vol .95, p .18-23.
[16] Yang, X., Lu, Y., Ma, Y., Li, Y., Du, F., and Chen, Y) .2006(, *Chemical Physics Letters*, Vol . 420, p .416-420.

IOP Conf. Series: Materials Science and Engineering **600** (2019) 012007 doi:10.1088/1757-899X/600/1/012007

Suppression of corrosion study on polypyrrole /zinc oxide nanoparticle composites for rechargeable battery electrode materials

Nicharee Thinnakornsutibutr[1, a], Teeranun Suracharoenchaikul[1, b], Pranut Potiyaraj[1, c] and Prasit Pattananuwat[1, 2, d]

[1]Department of Materials Science, Faculty of Science, Chulalongkorn University, Bangkok, 10330, Thailand
[2]Research Unit of Advanced Materials for Energy Storage, Chulalongkorn University, Bangkok, Thailand

[a]memee_nic@hotmail.com; [b]teeranun.sr@gmail.com; [c]pranut.p@chula.ac.th; [d]prasit.pat@chula.ac.th

Abstract. polypyrrole/zinc oxide nanoparticle (PPy/ZnO) composites are successfully synthesized via *in-situ* polymerization by aiding of ultrasonic irradiation. The different ratio of PPy to ammonium persulfate (APS) at 1:0.5, 1:1 and 1:1.5 is used to control the amounts of PPy in composites. Scanning Electron Microscopy indicates the development of nano/microstructure of PPy with increasing the amount of APS. The surface electrochemical activity of ZnO/PPy anodic electrode is characterized by linear polarization and cyclic voltammetry (CV). The introduce of PPy onto ZnO surface with the appropriate content and structure can improve the anti-corrosion and charge performance of electrode composites. The synthesized ZnO/PPy electrode at PPy/APS ratio of 1:1 can yield the highest values of corrosion inhibition efficiency of 73.50 % in 7 M KOH aqueous solution.

1. Introduction

Currently, several metals such as lithium, sodium and zinc are exploited as materials for battery electrodes. Among of them, zinc and its alloys are the most utilized materials as the anodic electrode for rechargeable aqueous batteries due to their high specific capacity, less expensive and safer for aqueous electrolyte [1, 2]. In addition, ZnO has been gained attraction for many researches owning to the modulating of its energy capacity with the different cathode materials such as manganese, nickel, vanadium and so on. Despite the its advantage, the zinc corrosion in alkaline electrolyte and the dendrite formation are crucial issues which directly affect to performance of zinc oxide electrode. Principally, the corrosion of ZnO electrode and the dendrite formation are occurred via the charge and discharge process. Zinc oxide is dissolved into the alkaline electrolyte and re-locates in non-uniform deposition on the electrode, inducing the dendrite growth on surface. Consequently, these issues lead to limit in the usages of rechargeable battery cycles and diminishing electrochemical properties.
To suppress the dendrite formation and the corrosion of zinc oxide anode, the surface modification of zinc oxide is necessary to reduce reaction rate of the dissolve/deposition. Several reports demonstrated the use of metallic oxides or carbon materials which can also improve the electrochemical performance of zinc oxide anode. However, those of metallic materials are low conductivity and the

IOP Conf. Series: Materials Science and Engineering **600** (2019) 012007 doi:10.1088/1757-899X/600/1/012007

cover coating on zinc oxide surface is difficult to control. Thus, the *coverable* materials with high electrochemical activity is need for surface modification. Recently, the conducting polymers (CP) such as polyaniline and polypyrrole are considered as conductive materials which can form tightly on nanoparticle surface [3, 4]. Yong et al. reported the using polyaniline coated onto zinc to inhibit corrosion reactions which can also decrease a self-discharge behavior and the hydrogen evolution reaction in batteries. Polypyrrole is one of interesting material together with high redox reaction. But, the use of PPy still has not been widely addressed for these issues [3].
Therefore, the aims of our study are to investigate the optimum condition for the synthesis PPy/ZnO composite and to compare the electrochemical performance with different amount of PPy coating at different amount of APS using, expecting that the appropriate PPy coated ZnO electrode will reduce the solubility of discharge process, prevent the dendrite formation of anode electrode and provide a good electrochemical behaviours.

2. Experimental

2.1. Preparation of polypyrrole coated ZnO nanoparticles

Polypyrrole coated zinc oxide nanoparticle was synthesized with ultrasonic polymerization. Briefly, 3 g of zinc oxide was dispersed in 20 ml of distilled water by 0.0966 g ammonium persulfate aqueous solution for 5 minutes. Then 0.1 mole of pyrrole in 12 M HCl was gradually added into the mixtures. The polymerization of pyrrole monomer was carried out under the ultrasonic irradiation at frequency of 280 W for 5 minutes, followed by filtration and washed with ethanol several times. The as-prepared PPy coated ZnO samples were obtained after drying at 60 ^{o}C overnight. In this present work, the different amount of PPy coated ZnO nanoparticles were prepared by modulating the mole ratio of PPy to APS at 1 to 0.5, 1 to 1 and 1 to 1.5 which were denoted as ZnO/PPy1, ZnO/PPy2 and ZnO/PPy3, respectively.

2.2. Fabrication of polypyrrole coated ZnO nanoparticle electrode

The sample electrodes were prepared by mixing of 80% wt of PPy coated ZnO, 15% wt of conductive carbon and 5 wt% polytetrafluoroethylene in water/ethanol solution. The mixture slurry was pasted on Ni foam and placed in oven at 60 ^{o}C for 3 h. The as-prepared electrode was cut in size 0.5 cm x 2 cm.

2.3. Characterizations

The characteristic functional groups of the samples were examined by Fourier transform infrared spectroscopy (FT-IR) with a Thermo scientific Nicolet 6700 (Tokyo, Japan). Polypyrrole coated zinc oxide nanoparticles were dried and coated with 10 nm of gold before being observed the surfaces using scanning electron microscopy (SEM: JSM-6480LV) at an acceleration voltage of 15 keV.

2.4. Electrochemical measurements

To evaluate evaluate the corrosion of the as-prepared ZnO coated PPy, linear polarization technique was applied to conventional three-electrode configuration with the as-prepared ZnO/PPy samples as working electrode, platinum plate as the counter electrode and Hg/HgO as the counter electrode by scanning between -0.75 and 0.35 V with its open circuit voltage (OCV) at the scan rate of 0.166 mV s^{-1}. The electrochemical performance was also determined by using cyclic voltammetry, employing in 7 M NaOH containing saturated ZnO between -0.1 to 0.6 V (*versus* Hg/HgO) at 1 mV sec^{-1}. To determine the corrosion inhibition efficiencies, the corrosion current density of anodic electrode (I_{corr}) was used to calculate the corrosion inhibition efficiency by equation 1.

$$\eta\ (\%) = \frac{\text{Icorr} - \text{I}'\text{corr}}{\text{Icorr}} \times 100 \qquad (1)$$

IOP Conf. Series: Materials Science and Engineering **600** (2019) 012007 doi:10.1088/1757-899X/600/1/012007

3. Results and discussion

In this work, Ppy/ZnO composites is *in-situ* polymerized under ultrasonic irrdaition at various Ppy/APS ratios. Representative surface morphologies of bare ZnO nanoparticles and PPy coated ZnO nanoparticles at different PPy/APS ratio are illustrated in fig. 1. SEM evidenance in fig. 1(a) reveals the typical structure ZnO with hexagonal prism-like nanostructure at average size of 50 nm. Fig. 1(b) exhibits the smooth morpology of PPy coated ZnO nanoparticles at PPy/APS ratio of 1 to 0.5. This morphology of PPy coated ZnO exhibits the similar morphology to bare ZnO, confirming that PPy can homogeneously coat onto ZnO nanoparticles surface. The more adding of APS directly affect to the efficiency of PPy coated onto ZnO surface. At high APS content, the presence of plate-like structure of PPy is found, as seen in Fig. 1c and d, indicating the rapid chain growth of PPy outside ZnO nanoparicles surface at high APS content.

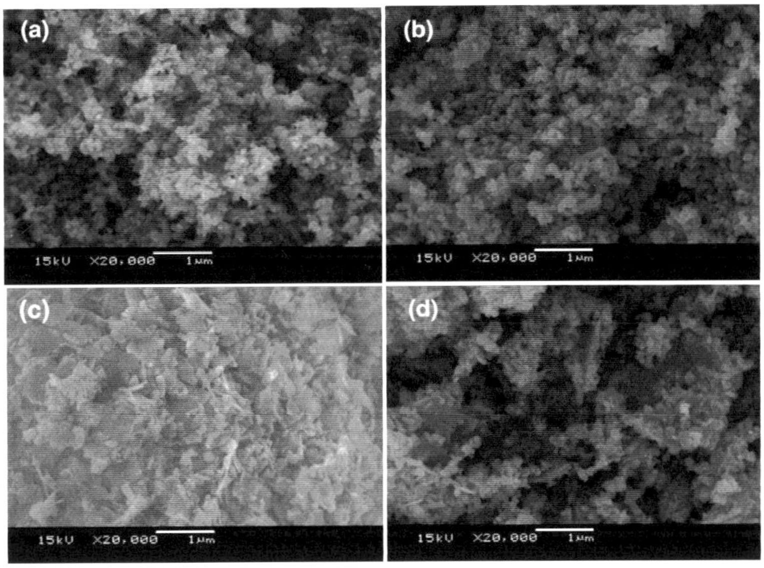

Figure 1. SEM images of bare zinc oxide and PPy coated zinc oxide at three different amounts of APS: (a) ZnO (b) ZnO/PPy1, (c) ZnO/PPy2 and ZnO/PPy3

Fig. 2 illustrates the FTIR spectra of the PPy coated zinc oxide samples at different PPy to APS ratio of 1.0 to 0.5, 1.0 to 1.0 and 1.0 to 1.5, respectively. All of PPy coated zinc oxide samples clearly reveal the evident characteristic of PPy signals. The broad peak at 3500-3300 cm^{-1} attributes to N-H stretching vibration. The peak at 1637 cm^{-1} is assigned to the characteristic peak for C=C stretching. The minor peak at 1270 cm^{-1} attributes to the C-N stretching of the pyrrole ring. The peak at 1165 and 1040 cm^{-1} corresponds to the C-H stretching of the pyrrole ring and the in-plane deformation vibration of N-H, respectively. The remarkable difference of the signal intensity of transmittance in each sample can be easily recognized, demonstrating that the more adding of APS in polymerization system leads to increase in PPy characteristic signals confirming the PPy chain growth is developed with increase APS content.

The linear polarization curves of the bare ZnO and PPy coated ZnO are illustrated in fig. 3. The calculated potential and current corrosion by Tafel fit are tabulated in table 1. Obviously, the bare ZnO electrode possesses the highest corrosion currents at 52.412 μA/cm2, while all PPy coated ZnO electrode reveal lower corrosion currents than that of ZnO, implying that the anti-corrosion performance of PPy coated ZnO is superior to ZnO. It can conclude that PPy coating on ZnO surface can suppress the corrosion of ZnO. The direct contact surface of ZnO with electrolyte is decreased by shielding effect of PPy coat, leading to slow down the ZnO corrosion process. ZnO/PPy2 reveals the lowest corrosion currents at 13.888 μA/cm2, confirming the best anti-corrosion performance. The slight increase in current corrosion at high APS content (ZnO/PPy3) is observed. This effect might be ascribed to the presence of the excess PPy coated ZnO and the inappropriate PPy structure, participating in the corrosion of PPy, the limitation of the ion mobility and the reaction channels. Consequently, the accumulation of charge on its surface is intensified resulting in increasing in the polarization [5].

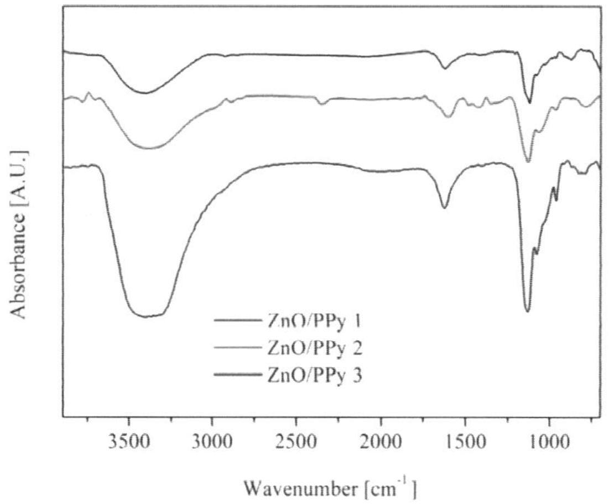

Figure 2. FTIR spectra of as-prepared PPy coated ZnO at different amounts of APS

Table1. Corrosion parameters for ZnO and PPy coated ZnO electrodes

Materials	Ecorr (V)	Icorr(μA/cm2)	η (%)
ZnO uncoated	-0.50967	52.412	-
ZnO/PPy1	-0.49904	20.138	61.58
ZnO/PPy2	-0.50483	13.888	73.50
ZnO/PPy3	-0.45277	15.207	70.98

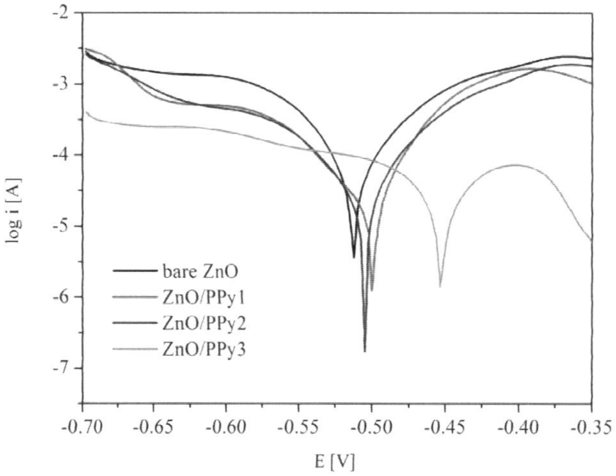

Figure 3. The polarization curves for bare ZnO and PPy coated ZnO electrodes

The electrochemical performance of ZnO and PPy coated ZnO electrode is investigated by cyclic voltammetry as seen in fig. 4. Undoubtedly, PPy coated ZnO affects the CV area by enhancing the electrical capacitive characteristic with increasing the PPy content. This result is due to the synergistic effect of PPy coated on ZnO nanostructure, inducing the electrical double layer and pseudo-capacitance behaviour of ZnO. However, at high PPy content, the dramatic decrease in electrical capacitive characteristics is observed for ZnO/PPy3. As evident support by SEM, the excess PPy chains reveal the inappropriate PPy microstructure. The ion mobility and the charge storage process are restricted, leading to increase electrode resistance and decrease electrical capacitive characteristic.

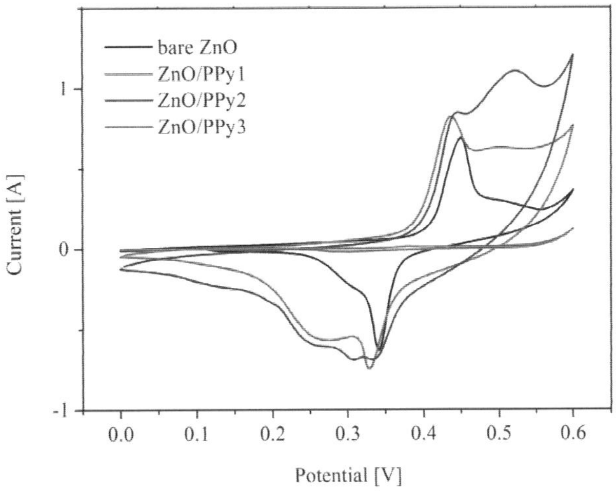

Figure 4. Cyclic voltammograms of bare ZnO and PPy coated ZnO electrodes

4. Summary

PPy/ZnO composites have been successfully synthesized at different PPy/APS ratio by aiding of ultrasonic irradiation. The use of very low concentration of APS as oxidizing agent exhibits the homogenous layer covering onto ZnO nanoparticle surface. Interestingly, the appropriate PPy features such a quantity and morphology can suppress the corrosion reaction in 7 M KOH medium. The synthesized ZnO/PPy electrode at PPy to APS ratio of 1:1 can yield the highest value of corrosion inhibition efficiency of 73.50 %. Evident results in CV manifestly demonstrates the better electrochemical active surface of PPy coated ZnO electrode at PPy to APS ratio of 1:1 relative to bare ZnO. This novel electrode is expected to utilize as an anode electrode in secondary ZnO battery for further study.

References

[1] A. Mishra, A. Mehta, S. Basu, S. J. Malode, N. P. Shetti, S. S. Shukla, M. N. Nadagouda and T. M. Aminabhavi: Mater. Sci. Ener. Tech. Vol. 1 (2018), p. 182–187
[2] M.A. Deyab: J. Power Sources Vol. 390 (2018), p. 176
[3] C. Cui, M. Li, X. Zhou, X. Zhang: Mater. Res. Bull. Vol. 112 (2019), p. 261
[4] Y. Jo, S. Kang, K. Prasanna, S. Eom, and C. Lee: Appl. Surf. Sci. Vol. 422 (2017), p. 406
[5] J. Huang, Z. Yang, Z. Feng, X. Xie, and X. Wen: Sci Rep. Vol. 6 (2016), p. 1

IOP Conf. Series: Materials Science and Engineering **600** (2019) 012008 doi:10.1088/1757-899X/600/1/012008

Preparation of MXene/N, S doped graphene electrode for supercapacitor application

Pusthira Kaewpijit[1], Jiaqian Qin[2, 4] and Prasit Pattananuwat[3, 4*]

[1]Program of Petrochemical and Polymer Science, Faculty of Science, Chulalongkorn University, Bangkok, 10330, Thailand
[2]Metallurgy and Materials Science Research Institute, Chulalongkorn University, Bangkok, 10330, Thailand
[3]Department of Materials Science, Faculty of Science, Chulalongkorn University, Bangkok, 10330, Thailand
[4]Research Unit of Advanced Materials for Energy Storage, Chulalongkorn University, Bangkok, Thailand

*prasit.pat@chula.ac.th

Abstract. In this study, novel MXene/N, S doped graphene electrode composites were sucessfully prepared via the simplistic drop casting process. The successful etching of Ti_3AlC_2 was evidently confirmed by SEM and XRD technique. The effect of N, S doped graphene adding at various ratio on electrochemical performance was exaimed by cyclic voltammetry and galvanic static charge-discharge, revealing that the intercalating structure of MXene-N, S doped graphene exhibits the synergistic effect to enhance the electrochemical properties. Especially, at MXene/N, S doped graphene ratio of 95 to 5, the highest specific capacitance of 599.7 F g^{-1} at 1 A g^{-1} is obtained, which is greater than that of the pristene MXene (185.8 F g^{-1} at 1 A g^{-1}) in 6M KOH.

1. Introduction

The recent discovery of two-dimension redox surface, namely MXene, has attracted attention from the scientific community for various applications including battery and supercapacitors because of its high redox properties and high surface area. The two-dimensional inorganic compounds combining of metallic conductivity of transition metal carbides and hydrophilic nature can yield the beneficial features of electrochemical performance. Technically, the two dimension structure features of MXene originated from Ti_3AlC_2 can be synthesised by etching process with hydrofluoric acid (HF) to remove Al layers [1]. However, several reports demonstrated that the restack of titanium layer can easily occur with irreversible process, leading to the loss of active surface area for redox reaction. Thus, the delamination of titanium layer is required to maintain 2D layer structure. According to the literature regarding the delamination process of MXene, the methodology including of the swelling process reveals the efficient method to expand the interlayer spacing, but the self-restacking still can occur under high pressure and temperature. Thus, it is still a challenge to find the suitable method to exfoliation the 2D features of MXene. Using intercalating materials has been proven for weakening the interactions between 2D layers. The insertion of small molecules such as activated carbon and carbon nanotube into titanium carbide layer is purposed to exfoliate and maintain the 2D structure of MXene. Among of carbon materials, graphene is an interested candidate for inserting into MXene layers. Furthermore, graphene is well known considering as electric double layer materials for supercapacitors application due to its high electrical conductivity and high surface area. Especially, graphene doped with nitrogen/sulfur atom possesses the outstanding electrochemical performance due to more electron donor structure. Thus, in this work, through the intercalation structure of MXene-N, S doped graphene is purposed to exfoliate/eliminate the self-restacking of titanium carbide layer and to improve the electrochemical activity.

2. Experimental

2.1. Synthesis of MXene via etching process

MXene powders were synthesized by etching of Al atom layers from the Ti_3AlC_2 precursors. Briefly, MXene powders were produced by immersing 0.5 g Ti_3AlC_2 in 10 mL 49% HF solution at 60 °C for 24 h. The obtained powders were washed, centrifugally separated and followed by drying in a vacuum oven at 70 °C for 24 h. Then, 0.5 g of the detached powders was combined with 10 ml of dimethyl sulfoxide (DMSO) for 24 h. Finally, the powders were rinsed with distilled water and drying in a vacuum oven at 60 °C overnight.

2.2. Synthesis of N, S doped graphene

Graphene oxide (GO) was prepared by a modified Hummers method from graphite powder. N, S doped graphene (N, S-G) was synthesized by refluxing of 2 mg/mL GO solution with thiourea solution at equivalent weight for 12 h. N, S-G was obtained after filtration and freeze-drying process.

2.3. Preparation of the MXene/N, S doped graphene electrode

MXene/ N, S-G samples were prepared at different ratio of 100:0, 99:1, 95:0 and 90:10, denoted as MXene, MXene99_ N, S-G1, MXene95_ N, S-G5 and MXene90_ N, S-G10, respectively. Briefly, the MXene95_ N, S-G5 was prepared by mixing of 0.01 g of MXene powder with 0.005 g of N, S-G in 20 mL dispersed water. Then the mixture was sonicated under ultra-sonication for 30 minutes at room temperature. The mixture was dropped onto the nickel foam sheet (2 cm x 1 cm) and dry at 60 °C.

2.4. Electrochemical measurements

The electrochemical properties of the as-prepared MXene/N, S-G electrodes were examined in 6M KOH using three-electrode system configurating with platinum as counter electrode and Ag/AgCl as reference electrodes. PGSTAT 12 equipment was used for cyclic voltammetry (CV) measurements and galvanostatic charge/discharge (GCD). CV experiments were executed in the range of 0 and 0.55 V at different scan rates. The range of 0-0.5 V potential at different current densities was adjusted for measuring the GCD curves.

2.5. Characterization

2.5.1. Fourier-transform infrared spectroscopy (FT-IR). FT-IR is used to identify the functional groups in the range of 500-4000 cm^{-1} fabricating with KBr.

2.5.2. X-ray diffractometer (XRD). XRD is used to determine the interlayer spacing between Ti_3AlC_2 and MXene layer. XRD patterns were obtained on a Bruker XRD using Cu Kα radiation (λ = 0.15406 nm) with a scan speed of 2° min^{-1} at wide-angle range (2θ value 8 to 60°), the accelerating voltage of 40 kV and the emission current of 40 mA. The positive integer (n) is 1.

3. Results and discussion

XRD is used to analyze of exfoliation process to synthesis MXene. Fig. 1 presents the XRD pattern of Ti_3AlC_2 before (1a) and after etching with HF (2b). The starting Ti_3AlC_2 reveals the typical XRD quite

IOP Conf. Series: Materials Science and Engineering **600** (2019) 012008 doi:10.1088/1757-899X/600/1/012008

pure Ti_3AlC_2 [2]. Meanwhile, after exfoliation process by HF, the most intense peak at 2θ = 39.5° corresponding to the plane (104) is disappeared. The (002) peak is broadened and shifted to lower angles from 2θ = 9.7° (d_{002}=0.91 nm) to 8.7° (d_{002}=1.02 nm), indicating that the successful extraction of Al layer and the loss of its original layered structure after etching process [3]. In addition, XRD pattern of anatase TiO_2 at 2θ = 25.3°, 37.9°, 48.0° and 53.6° is found corresponding to (101), (004), (002), and (105) planes, respectively [4], implying that MXene has been partially oxidized during the annealing and etching process. [5]. The impurity of TiC are also indexed at 2θ = 36.4° (111) and 42.3° (200) [6].

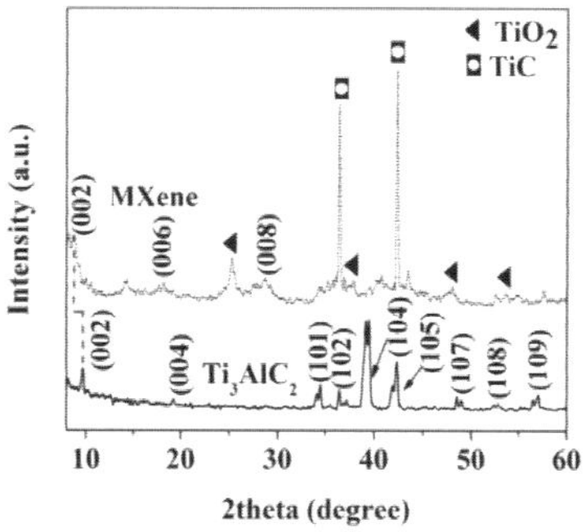

Figure 1. XRD patterns of (a) Ti_3AlC_2 and (b) MXene

The morphology of the Ti_3AlC_2 and MXene are investigated by SEM technique. Typical SEM images of the Ti_3AlC_2 and the exfoliated MXene nanosheets are shown in fig. 2(a) and 2(b), respectively. As seen in fig. 2(b), after HF-etching, the brick morphology of Ti_3AlC_2 is obviously transform to accordion-like morphology of Ti_3C_2, confirming that Al atom in Ti_3AlC_2 phase are selectively etched [7]. Fig. 2(d) shows the homogenous surface of MXene95_N, S-G5 in micro/nano structure on nickel foam.

Figure 2. Typical SEM images of the (a) Ti_3AlC_2, (b) MXene, (c) Nickel foam (d) Nickel foam coated MXene95_N, S-G5

FT-IR spectrum of graphite, graphite oxide and N, S-G materials are shown in fig. 3. It is clear that no distinct peak can be found from the graphite spectra. Graphite oxide exhibits representative peaks at 3270, 1714, 1606, 1389 and 1070 cm^{-1} corresponding to O-H stretch, C=O stretch, aromatic C=C, C-O stretch, respectively [8]. The FT-IR spectra of N, S-G reveals the peak at 3145 cm^{-1} and 3433 cm^{-1} attribuing to N-H and O-H stretching vibrations, respectively. These considerable amino and hydroxyl groups can enable hydrophilicity. The peak at around 1728 cm^{-1} corresponds to the stretch vibration of C=O carboxylic groups and the peak at 1375 cm^{-1} can be assigned to the vibration of C-N. Furthermore, the existence of C-O and S=O bonds on N, S-G is confirmed by the peak of 1211 and 1110 cm^{-1}, respectively [9].

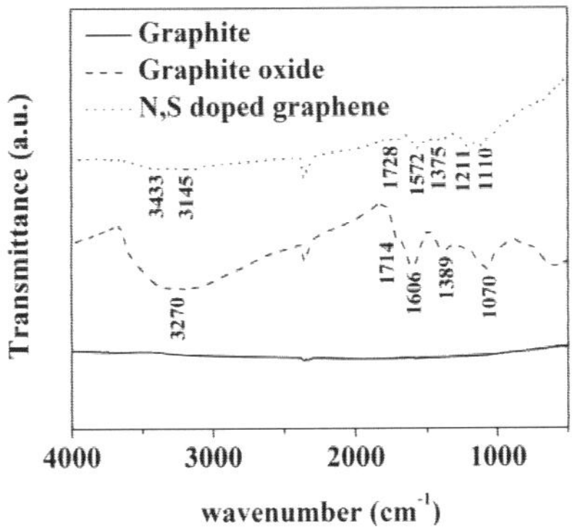

Figure 3. FT-IR spectra of graphite, graphite oxide and N, S doped graphene.

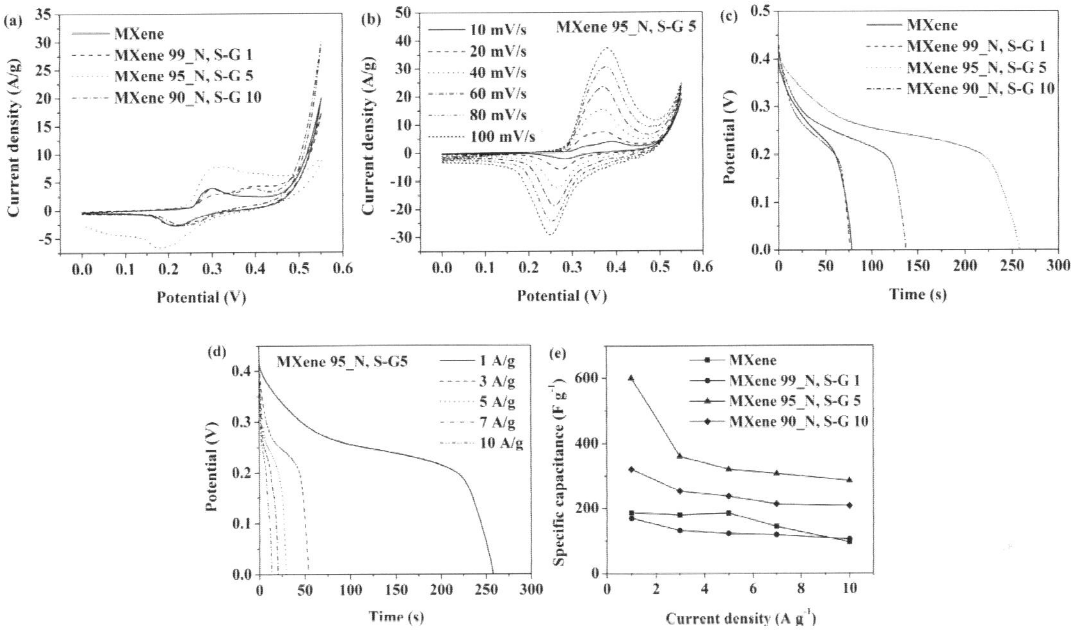

Figure 4. (a) CV curves of the as-prepared samples at 10 mV s^{-1}, (b) CV curves of MXene95_N,S-G5 electrode at different scan rates, (c) GCD profiles of the as-prepared samples at 1 A g^{-1}, (d) GCD profiles of MXene95_N, S-G5 electrode at different current densities and (e) the calculated specific capacitance by GCD of the as-prepared samples

Fig. 4(a) shows the CV curves of the MXene and Mxene/N, S-G electrodes at different ratio (MXene99_N, S-G1, MXene95_N, S-G5 and MXene90_N, S-G10) at 10 mV s^{-1} in 6M KOH. MXene, MXene90/N, S-G10 and MXene99/N, S-G electrodes show similar CV areas. Remarkably, The largest CV area is MXene95_N, S-G5 implying the highest electrochemical performance. In addition, fig. 4 (b) demonstrates CV curves of MXene95_N, S-G5 electrode with different scan rates from 10 to 100 mV s^{-1}. A pair of redox peaks of samples was clearly observed for each curve within 0-0.55 V and can be mainained, even at high scan rate, indicating that the capacitance mainly comes from the pseudocapacitance associated with reversible intercalation/deintercalation of protons along with the change of Ti oxidation state [10]. GCD profiles of MXene and Mxene/N, S-G electrodes at 1 A g^{-1} is revealed as seen in fig. 4(c), demonstating the longest discharge time of MXene95_N, S-G5. These results are also corresponded with CV results, implying that the MXene95_N, S-G5 can yield the highest specific capacitance. GCD curves of MXene95_N, S-G5 at various current densities are shown

IOP Conf. Series: Materials Science and Engineering **600** (2019) 012008 doi:10.1088/1757-899X/600/1/012008

in Fig. 4(d) revealing that the electrochemical performance can maintain 52.6% with the increase current from 1 to 10 A g^{-1}. The calculated specific capacitance values by GCD for all electrode samples at the different proportion of N, S-G are presented in fig. 4(e). Among of these electrodes, MXene95_N, S-G5 electrode exhibits the highest specific capacitance of 599.7 F g^{-1} at a current density of 1 A g^{-1}, slightly higher than those of pristine MXene (185.8 F g^{-1}), MXene99_N, S-G1 (169.0 F g^{-1}) and MXene90_N, S-G10 (319.9 F g^{-1}). Clearly, MXene95/N, S-G5 yield the optimum condition.

4. Summary

In summary, MXene/N, S-G electrodes were successfully synthesized through drop casting method. The experiments indicate the highest specific capacitance of MXene/N, S-G electrode is MXene95_N, S-G5 (599.7 F g^{-1}) at 1 A g-1 which is arisen from the intercalation structure of MXene-N, S doped graphene exhibiting the synergistic effect to enhance the electrochemical properties. Evidence results confirmed that MXene/N, S-G electrodes are promised material for the supercapacitor application in the future.

Acknowledgments

The work is executed at financial support of Research and Development of Alternative Energy Technology Project (Budget Year 2018, 2nd Round) Energy Conservation and Promotion Fund.

References

[1] B. Anasori, M. R. Lukatskaya and Y. Gogotsi: Nat. Rev. Mater. Vol. 2 (2017)
[2] P .Yan, R. Zhang, J. Jia, C. Wu, A. Zhou, J. Xu and X. Zhang: J. Power Sources. Vol. 284 (2015), p. 38-43
[3] G. Lv, J. Wang, Z. Shi and L. Fan: Mater. Lett. Vol. 219 (2018), p. 45-50
[4] H. Wang, R. Peng, Z. D. Hood, M. Naguib, S. P. Adhikari and Z. Wu: Wiley-VCH
[5] Q. X. Xia, J. Fu, M. Yun, S. Mane and K. Ho: RSC Adv. Vol. 7 (2017), p. 11000–11011
[6] W. Sen, H. Sun, B. Yang, B. Xu, W.Ma, D. Liu and Y. Dai: Int. J. Refract. Met. Hard Mater. Vol.28 (2010), p. 628-632
[7] S. Xu, G. Wei, J. Li, Y. Ji, N. Klyui,V. Izotov and W. Han: Chem. Eng. J. Vol. 317 (2017), p. 1026–1036
[8] J. C. Yan, X. Q. Zeng, T. H. Ren and E. van der Heide: Biosurf. Biotribol. Vol. 1 (2015), p. 113-123
[9] P. Guo, F. Xiao, Q. Liu, H. Liu, Y. Guo, J. R. Gong, S.Wang and Y. Liu: Sci. Rep. Vol. 3 (2013) p. 1-6
[10] J. Yan, C. E. Ren, K. Maleski, C. B. Hatter, B. Anasori, P. Urbankowski, A. Sarycheva and Y. Gogotsi: Adv. Funct. Mater. Vol. 1701264 (2017), p. 1–10

IOP Conf. Series: Materials Science and Engineering **600** (2019) 012009 doi:10.1088/1757-899X/600/1/012009

Preparation of engineered stones

Saowanee Suta[1,2], Suthee Wattanasiriwech[1,2*], Darunee Wattanasiriwech[1,2] ,Sitthi Duangphet [1,2], Chuleeporn Thanomsilp[2]

[1]Center of Innovative Materials for Sustainability, Mae Fah Luang University, Chiangrai, Thailand
[2]School of Science, Mae Fah Luang University, Chiangrai, Thailand

Corresponding author: suthee@mfu.ac.th; darunee@mfu.ac.th

Abstract. Engineered Stones were produced in a laboratory using polyester resin and domestic quart stones. Selection of these particle sizes were based on packing theory and simple Stroke's law. From combination of both scenario, three different sizes of quartz were calculated and selected. In this scenario, the smallest size was calculated to be 6 μm while the other size ranges were calculated based on the 7x factor. To test the idea, another two sets of samples were also prepared for comparison. The samples were fabricated by vibration or uniaxial pressing. It was found that all the sample sets fabricated by pressing offered better density and Vickers hardness due to the better packing and higher solid loading of the composite. However, the sample set which strictly followed the calculation did not provide the best density and hardness suggesting that the simple model may not be fully correct and some missing parameters may be required. The best sample set belonged to one which followed 7x factor but with a larger starting quartz size. Sample set with the best properties was comparable to the commercial products.

1. Introduction

Natural stones have been used as decorative construction materials due to their aesthetic qualification, accessibility and strength [1]. However, they are often regarded as hard but fragile so combination with other materials to adequately offer more desirable qualities have been addressed [2, 3]. This new composite material is quoted as "engineered stone or "agglomerated stone" as according to BS EN 14618:2009. Engineered stone is thus defined as a composite material made from stone particles packed together by polymer resin or cement to obtain mechanical properties closed or superior to natural stones. Engineered stone typically contain 90-94% natural stones and other minerals [3].
Applications of engineered stones in construction industry may be ranged from exterior cladding to table or bench slabs for interior use applications. For use as exterior cladding, properties such as light weight, low thermal conductivity was achieved by using aggregate of low density was reported [4]. In this patent, crushed limestone, silica sand, and ground calcium carbonate were used as the aggregate while the unsaturated polyester resin binder was improved by adding 1:1 mix of polypropylene (PP) and polyethylene (PE), and about 1% methyl ethyl ketone peroxide catalyst. However, the product showed low thermal resistance and low resistance to staining. Low porosity engineered stones offered a variety of applications especially those required high mechanical properties, surface maintenance free and resistance to chemical attacks. These products found their use in diverse places such restaurants, homes, hospitals, laboratories, and in many more, so different sets of thermal properties

might be needed for different purpose of use. The work in *ref* 3 showed that when heated up to 100 °C, commercial engineered stones showed superior flexural strength than natural lime stone and granite. After being submitted to rapid cooling rom 200 °C down to 20 °C, engineered stones continued to show higher values of flexural strength as compared to the natural stones.

Various processing methods with have been discussed in literatures [4, 5]. Compaction vibration in a vacuum environment was used to compact engineered stone slab whose main solid components were waste glass and fine granite aggregate. The final product showed high compressive strength of 148.8 MPa, water absorption < 0.02%, and flexural strength of 51.1 MPa which were superior to those of natural construction slabs [5]. The research results revealed that vibration was the key processing step required to adjust the orientation of aggregates to become more closely compacted so densified product could be obtained. Only pressing failed to give the slabs with so desirable properties. A similar method so-called "vibro-compression under vacuum" was patented in the USA [US 7,815,827 B2]. Autoclave polymerization of polymeric resins mixed with stone gravels for production of synthetic marble slab was patented in 2014 [US 2012/0196087 A1]. In the study by Khater and Ezzat [2018], engineered stone was prepared using geopolymer resin instead of polymer resin [2].

To obtain minimal porosity, packing of gravel particles used is essential. In packing of monosize spherical particles, the highest density could be obtained was only 74%. With irregular shaped-particles, packing of monosize particles would bring down the number to only 56% [6]. To achieve higher packing density, particles of multi-sizes must be employed. However, according to Stroke's law preparation of engineered stones is performed in liquid polymer resin, where dragging force acting on particles depends on viscosity of the liquid, density and particle size of the solids [7]. In this research, particle packing was designed and quartz particles of mixed sizes were selected and used for preparation of engineered stones. Size selection was based on Stroke's law. Density and hardness of the slab was the main focus of our present study.

2. Experiments

2.1. Packing design of particles

Size selection for particle packing was designed based on particle packing theory and also based on Stroke's law [7]. In packing theory, triangular packing of iso-spherical size particles would leave an interstitial site with the size ratio of about 0.155 or 7 times smaller than the large particles. This is the smallest size that can fit into the interparticle voids and was selected following Stroke's law. From Stroke's law, the resistant force (F) acting on a falling particle in a liquid as it is sinking under the influence of gravity is a function of its size and viscosity of the liquid as follows;

$$F = \rho V g = 6\pi r \eta v \tag{1}$$

where r is radius of the spherical particle, η is viscosity of the liquid, v is velocity of the sinking particle, ρ_S is density of the solid while ρ_L is density of the resin, V is volume of the particle which is equal to $\frac{4}{3}\pi r^3$. Eq. 1 can then be re-written as:

$$r = 3 \times \sqrt{\frac{\eta v}{2(\rho_S - \rho_L)g}} \tag{2}$$

Where η was measured viscosity of the resin while velocity of the sinking particles was calculated based on the falling distance and the curing time of the resin.

From Eqn. 2, size of the smallest particles was thus calculated to 5.18 μm. It must be kept in mind that there are parameters omitted in this simple model. Those parameters are such as variation of viscosity of the resin with time, true morphology of the particles, Brownian motion. The packing of 3 sized-particles was designed to be 3 sets as follows;

Set 1: The biggest size was chosen to be in the range of 500-1,000 μm while the medium and smallest particles were selected to be smaller by 2-5x factor so they were 250-500 μm and 75 μm respectively. This sample set was used as a compared set.
Set 2: The biggest size was chosen to be in the range of 500-1,000 μm, while the medium and smallest particles were selected in accordance with the 7x factor. The smallest particles were about 16 μm in size while the medium and large particles were in the size range of 108-125 μm. This sample set was used as a compared set.
Set 3: The smallest particles were about 6 μm in size while the medium and large particles were, as in accordance with the 7x factor, in the range of 37-44 μm and 250-350 μm respectively. This sample set follows the calculation's guidelines exactly in both size ratio factor and the starting size range.
To obtain the required size, domestic quartz stones were crushed, ground and finally sieved through metal sieves.

2.2. Processing and characterisation
Polyester resin was first mixed with a hardener prior to addition of the ground quartz. Two shaping methods, vibration casting and compression, were compared so the mixture recipe was selected appropriated to the method. For the vibration casting, quartz to polymer resin ratio was 75 to 25 by weight. The mixture was vibrated at 25 Hz for 2 min before being cured at 80 °C for 30 min. In another shaping method, the mixture was pressed at 5 MPa for 1 min before being cured at the same condition. The mixture ratio was selected to be 90 to 10 by weight. The samples shaped by the vibrational method was coded as Set x-a while the samples shaped by pressing were coded as Set x-b.
Apparent density was measured following the International Organization for Standard (ISO 10545-3) as shown in Eq 3 [8].

$$\rho = \frac{m_{dry}}{m_{dry} - m_{susp}} \quad (3)$$

where m_{dry} is weight of the sample dried at in oven at 110±5 °C for 24 hr, m_{sus} is the suspended mass in water. The ideal density of the engineered stone was calculated based on the composition of the composite with quart and resin density of 2.65 and 1.20 g/cm^3 respectively. The calculated ideal density of Set x-a was 2.28 g/cm3 while that of Set x-b would be 2.50 g/cm^3.
Vickers hardness was measured using micro-hardness. A square-based pyramid indenter whose opposite sides meet at the apex at an angle of 136° was employed. Hardness in Vickers was calculated using a formula in Eq 4;

$$HV = 1.854\,\frac{F}{d^2} \quad (4)$$

where d is diagonal length of the indented surface and F is the applied load.
A scanning electron microscope (SEM) LEO/1450 VP at an accelerating voltage of 10 kV was used for microscopical observation.

3. Results and discussion
Apparent density of the Set x-a and Set x-b samples are graphically displayed in Fig. 1. The calculated ideal density for Set x-a is 2.28 g/cm^3 while that of Set x-b is 2.50 g/cm^3. In general samples of the same composition in Set x-a had slightly lower density than those in Set x-b. The maximum density was found in both Set 2-a and –b suggesting the best packing of the particles. From our hypothesis, the best packing of moving particles according to both closest packing principle and Stroke's law should belong to Set 3. However, our results showed that Set 2-a and -b whose packing was based on 7x ratio, showed the best packing efficiency amongst the samples of the same shaping method. Set 1 showed relatively similar or slightly higher density than Set 3.

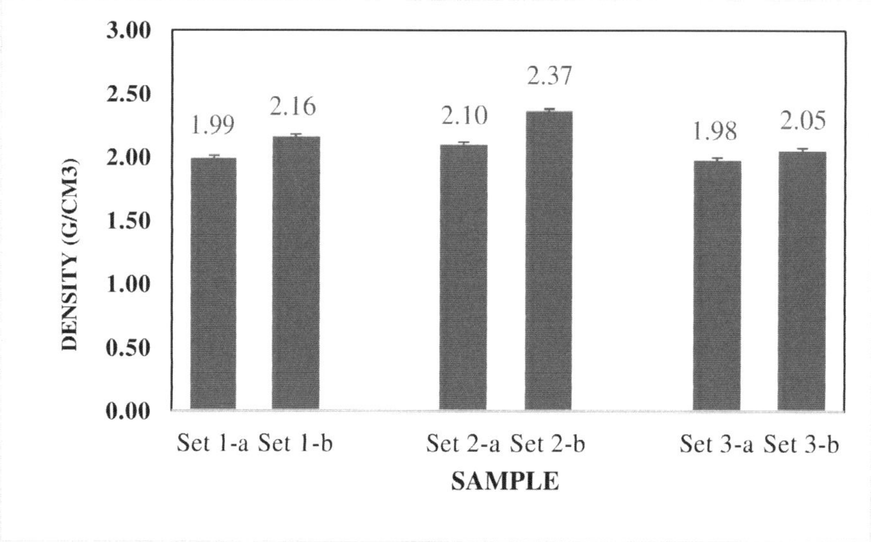

Figure 1. Apparent density of the two quartz stone sample sets.

Vickers hardness results are shown in Fig 2. Vickers hardness results for Set 1-a and –b and Set 2-a and –b showed a similar trend to the density i.e. the higher the density, the higher the hardness. Set 3-a, however, showed the poorest hardness, possibly due to the deeper sinking of particles below the resin surface. It was noted that Set 3-b sample was too crumble so measuring hardness was not possible. In good agreement with the density results, samples in Set x-b showed better hardness than those in Set x-a. The best Vickers harness was found in Set2-b due to its best packing efficiency according to the related density result. To compare the result, some commercial products were selected and tested. The average density and Vicker hardness were 2.31±0.01 g/cm^3 and 833.21±27.10 HV.

Microstructure of the Set X-a and Set X-b samples are shown in Fig 3 and 4 respectively. Set 1-a (Fig. 3a) showed relatively larger space between particles while space filling with particles of 3 different sizes seems to be more effective in Set 2-a (Fig. 3b). Space filling in Set 3-a (Fig. 4c) was mainly occurred by two size of particles. Set 1-b (Fig. 4a) sample showed similar particle arrangement similar to Set 1-a but with a better space filling scenario. Set 2-b (Fig. 4b) showed existence of 3 particle sizes with a relatively smaller distance between particles. This sample showed the greatest density in good agreement with its microstructure.

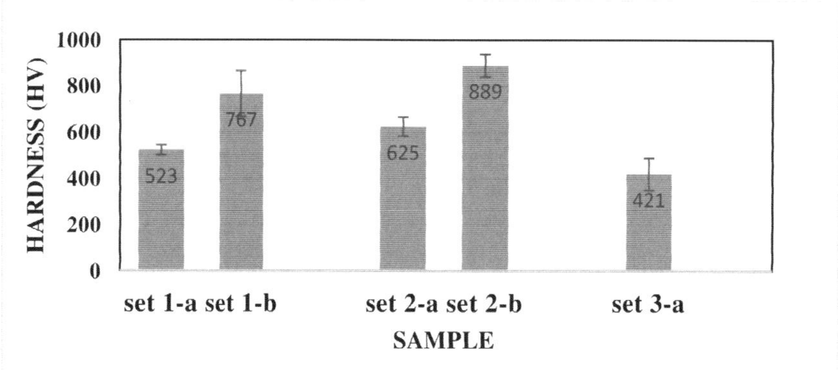

Figure 2. Vickers hardness results of the quartz stone samples sets. Noted that Set3-b was too crumble so measuring hardness was not possible.

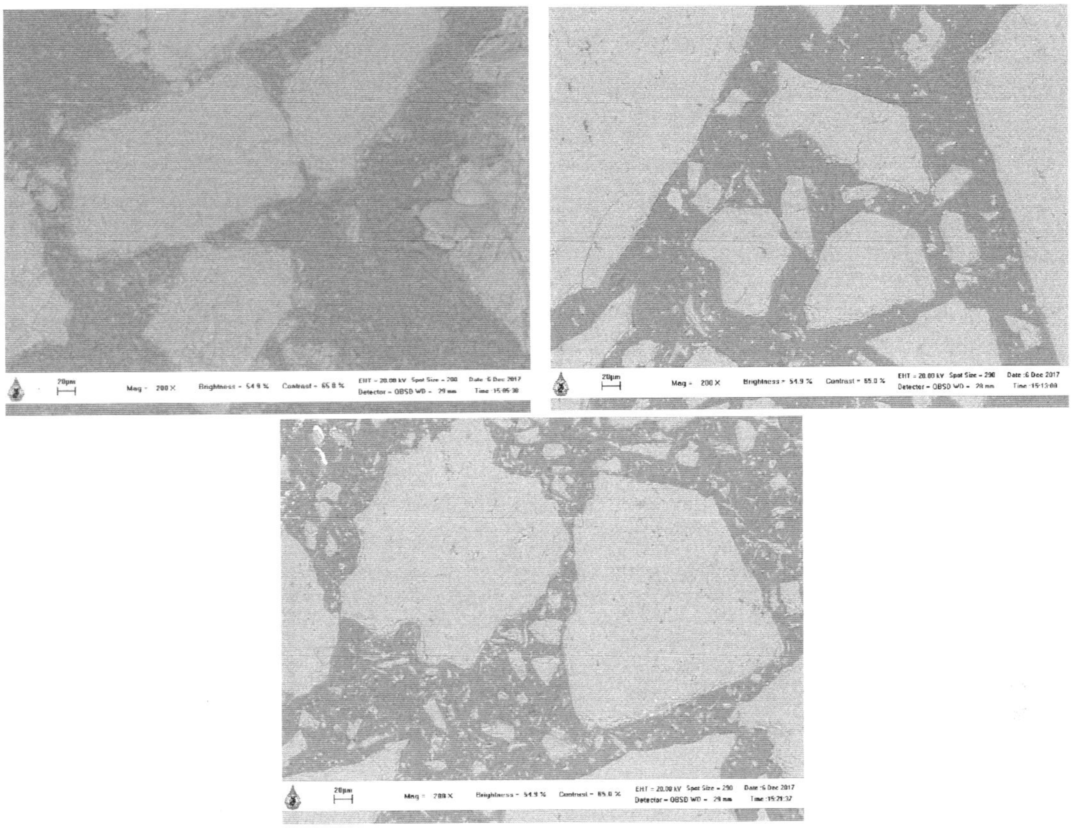

Figure 3. Back scattered electron microscopical pictures for (a) Set 1-a, (b) Set 2-a and (c) Set 3-a.

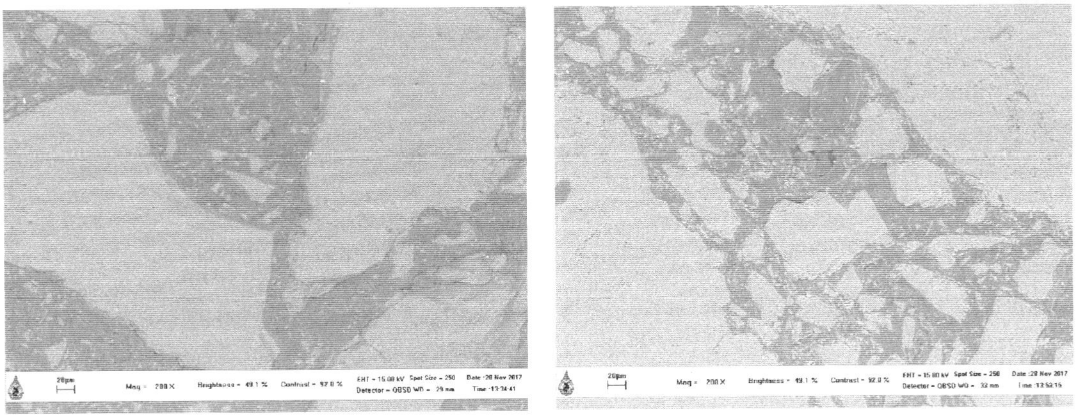

Figure 4. Back scattered electron microscopical pictures for (a) Set 1-b and (b) Set 2-b.

4. Conclusion

From our results, it was suggested the closest packing model gave satisfactory result as seen in both Set 2-a and –b. Set 3-a and –b had adopted the closest packing model along with the simple Stroke's law for the drag force. This model was, however, cancelled off many parameters so in practical it may not be best fit. Practical parameters such as variation of viscosity of the resin with time, morphology of the particles, Brownian motion, and the falling distance may be needed to put together in calculation, though this will make it extremely complex.

Acknowledgement
This project is granted by the Thailand Talent Mobility 2016 Project. Financial support from Mae Fah Luang University for Conference attendance support is highly appreciated.

References
[1] S. Hamoush, T. Abu-Lebdeh, M. Picornell, S. Amer, Development of sustainable engineered stone cladding for toughness, durability, and energy conservation, Construction and Building Materials 25 (2011) 4006–4016.
[2] H.M. Khater, M. Ezzat, Preparation and characterization of engineered stones based geopolymer composites, Journal of Building Engineering 20 (2018) 493–500.
[3] J.P. Lam dos Santos, L.G. Rosa, P.M. Amaral, Temperature effects on mechanical behavior of engineered stones, Construction and Building Materials 25 (2011) 171–174.
[4] K. Wiener Cultured stone method of manufacture. US Patent # 6056904; February 12, 1998.
[5] M.Y. Lee, C.H. Ko, F. C. Chang, S.L. Lo, J.D. Lin, M.Y. Shan, J. Lee, Artificial stone slab production using waste glass, stone fragments and vacuum vibratory compaction, Cement & Concrete Composites 30 (2008) 583–587.
[6] H. Groemer, Some basic properties of packing and covering constants, Discrete and Computational Geometry 1(2) (1986), 183-193.
[7] G.K.Batchelor, *An Introduction to Fluid Dynamics*. (1967). Cambridge University Press, pp 131-173.
[8] The International Organization for Standard (1995), Determination of water absorption, apparent porosity, apparent relative density and bulk density.

IOP Conf. Series: Materials Science and Engineering **600** (2019) 012010 doi:10.1088/1757-899X/600/1/012010

Preparation of graphene /poly (diallyldimethylammonium chloride) – poly (styrene sulfonate) polyelectrolyte composites and their electrochemical performance

Prapassorn Areebumrungsuk[1, a], **Pronteera Sukhsuwan**[1, b] **and Prasit Pattananuwat**[1, 2, c]

[1]Department of Materials Science, Faculty of Science, Chulalongkorn University, Bangkok, 10330, Thailand
[2]Research Unit of Advanced Materials for Energy Storage, Chulalongkorn University, Bangkok, Thailand

[a]A.Prapass@gmail.com; [b]pronteera.sukh@gmail.com; [c]prasit.pat@chula.ac.th

Abstract. Here, we successfully prepare graphene /poly (diallyldimethylammonium chloride) – poly (styrene sulfonate) (G/PDADMAC-PSS) polyelectrolyte composites by simple mixing method. The effect of NaCl adding at 2, 4, 6, 8 and 10 % wt. on the sizing of polyelectrolyte composites (PECs) is investigated. The concentration of NaCl ion plays an important role in size control by the formation of the primary polyelectrolyte composites and following by the agglomeration itself. The electrochemical properties are evaluated by cyclic voltammetry demonstrating that G/PDADMAC-PSS PECs at 8 %wt. can yield the highest specific capacitance of 68 m $F.cm^{-2}$ at 10 mV sec^{-1}

1. Introduction

Polyelectrolyte complexes are formed by the association of oppositely charged polymers, *assembling* electrostatic interaction between polycations and polyanions. The direct mixing of dispersed aqueous solutions of oppositely charged PECs can formed a dense phase of the nano-polymer molecule that is separated from the medium [1]. Moreover, the modulating of ionic interaction by salt ion addition can customize the shrinking/extending of the PECs molecule. This class of PECs polymer is practically utilized for reproducible preparation and controllable sizing. These unique characteristics are key requirements to improve the nanoparticle features, especially addressing for biomedical and pharmaceutical application. One of interesting that still has not been sufficiently addressed is to introduce the other substances such as carbon nanomaterials in PECs. The nanoparticles in the core shell surrounding with PECs can offer the charge surface with controlling sizing the nanoparticles. [2]
Graphene is a form of nanosheet carbon which has a single layer of hexagonal lattice structure, high electrical, thermal conductivity and high surface area. Principally, graphene is known as the forefront materials for energy-storage research and widely used for EDLC owing to its excellent electrical conductivity and high surface area [3,4]. However, the agglomeration of graphene during fabrication causing the decrease in surface area is critical issue to obtain the high electrochemical performance.[5]
Therefore, the aim of this work is to study the characteristics of the introduce of graphene into poly (diallyldimethylammonium chloride) - poly (styrene sulfonate) polyelectrolyte complex systems. The

effect of salt ion concentration on the customizing size of G/PDADMAC-PSS PECs is purposed. Additionally, the efficiency of electric charge of PECs on graphene is evaluated by cyclic voltammetry [6]. This G/PEC material is expected to fulfill the novel class material electrodes composites, utilizing on the surface materials of any shape.

2. Experimental

2.1. Materials
Graphite powder (> 99% purity, Aldrich), Poly (diallyldimethylammoniumchloride) (20 %wt., MW 200,000 - 350,000 g/mole, Aldrich) and Poly (sodium 4-styrenesulfonate) (MW 70,000 g/mole, > 99% purity Aldrich) were used as received. Other the reagents were of analytical grade and used as received without any further purification.

2.2. Preparation of graphene
GO was synthesized by the modified Hummers' our modified method. Briefly, 6.0 g of graphite and 3.0 g of $NaNO_3$ was added into 150 ml of H_2SO_4, followed by adding the first portion of 18.0 g $KMnO_4$. The reaction was kept stirring at 35 °C for 7 h. After that, the second portion of $KMnO_4$ (18.0 g) was added and employed at 50 °C for 7 h. Then, the reaction mixture was poured into ice water, followed by adding H_2O_2. The slug products were washed with dilute HCl and further neutralized with deionized water. Graphite powder was obtained after dry at 60 °C. Graphite powder was further dispersed in distilled water by aiding of ultrasonic and chemically reduced with sodium ascorbate at 60 °C for 12 h. The obtained reduced graphene was further washed with deionized water several times and freeze dry overnight.

2.3. Preparation of graphene/Poly(diallyldimethylammonium chloride) - Poly(styrene sulfonate) Polyelectrolyte composites
6.0 mg of graphene was dispersed in 60 ml of 0.1 M PSS aqueous solution by aiding of ultrasonic irradiation for 3 h. Then, the solution was mixed with 30 ml of 0.1 M PDADMAC solution at 25 °C for 30 minutes. The various of NaCl content (2, 4, 6, 8 and 10 %wt.) was added into the mixture and then kept stirring at 25 °C for 30 minutes to obtain the polyelectrolyte composites.

2.4. Characterizations
The characteristic functional groups of the graphene oxide and graphene were examined by Fourier transform infrared spectroscopy (FT-IR) with a Thermo scientific Nicolet 6700 (Tokyo, Japan). The laser diffraction particle size analyzer (SALD-7101) was utilized to determine the size distribution of the G/PDADMAC/PSS samples in different NaCl concentration.

2.5. Electrochemical measurements
The electrode samples were prepared by direct pasting the G/PDADMAC/PSS samples onto carbon fiber paper without the additional of binder. This paste active material can easily apply on any surface materials. The electrochemical performance was determined with three electrode consisting of the as-prepared samples as working electrode, platinum as the counter electrode and Ag/AgCl as the reference electrode by using cyclic voltammetry, employing in 50 ml of 1 M NaCl between 0 to 1 V from 10 to 100 mV/sec.

3. Results and discussion
FTIR spectra of graphite, synthesized GO and G are exhibited as seen in Fig. 1a. There is no significant signal for graphite spectrum due to non functional group of graphite. For GO, the oxidized structure of graphite is examined. The broad center peak at 3400 cm^{-1} attributes to OH stretching vibration of adsorbed water and -OH group functional. The peaks at 1728, 1628 and 1388 cm^{-1} are assigned to C=O, C=C and C-O stretching vibrations, respectively. In contrast, G reveals the dramatic

decrease in oxygen functional groups of GO which occurs from redox reaction, mainly consisting of decrease at the intensities peaks of 1728 (C=O) and 1388 (C-O) cm^{-1}.

XRD spectra is used to confirm the planar characteristic of the graphite, synthesized GO and G (Fig. 1b). Graphite exhibits the typical crystal structure of graphite at 26.5°. The calculated intercellular space by Bragg's Law is 3.36 Å. After oxidation process, GO displays the shift of (002) position to the sharp peak at 10.8°, implying that the space between layer is increased to 8.16 Å. The reduced graphene oxide shows the board peak at 25.8° with the disappearance of the sharp peak, confirming the exfoliation of layered structures of graphene.

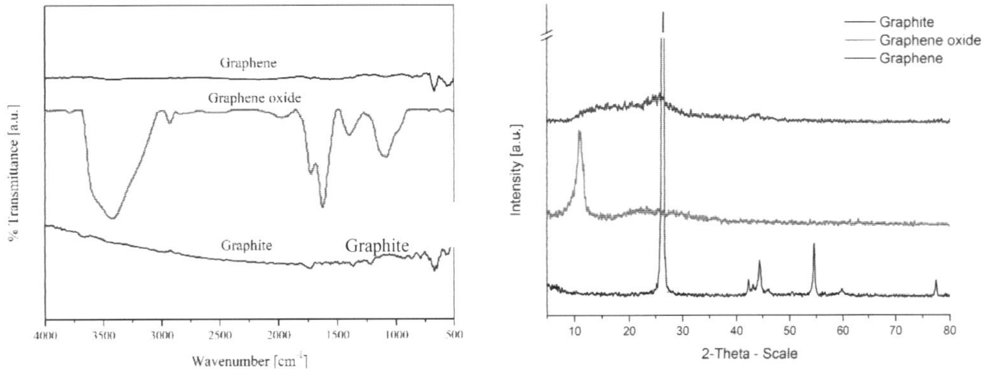

Figure 1. (a) FT-IR spectra of GO and (b) XRD spectra of GO and G

In this work, the polyelectrolyte complexes is formed by the direct mixing of the G/PDADMAC and PSS solution at different NaCl concentrations. As evident in Fig. 2a, the formation of the aggerated G/PDADMAC-PSS PEC manifestly reveals the size dependence on NaCl concentration. The formation of the aggregated PEC particles in molecular level is arose from the result of the agglomeration of the primary colloid particles, when the PEL charge is disturbed by changing of Na^+ and Cl^- concentration in Fig. 2b [1]. The PSD of the primary G/PEC particle molecule from mixed PEL solutions is examined by laser diffraction particle size analyzer. The aggregated G/PEC particles is diluted and centrifuged to separate the G/primary colloid particles from mixed PEL solution. PSD result in Fig. 3 demonstrating that the primary G/PEC at 2 %wt. NaCl in aqueous medium exhibits a wide size distribution with an effective particle size of 50 μm, which is almost nearly to the graphene size (33 μm), indicating the high efficiency of PEC surrounding on graphene surface in nano to submicrometer level. Principally, the addition of NaCl after mixing PEC affects rather outer particle surface than the inner particle. The adding NaCl up to 6 %wt. leads to swell in primary PEC particles size on graphene surface (average particle size of 600 μm). At 8 %wt. of NaCl adding, the primary PEC on G surface starts to shrink with the major average size of 80 μm and minor average size of 600 μm. This results might be explained by the changing of the conformation of PECs molecules at the critical salt concentration. The rapid aggregation process is initiated and the large aggregated PEC is formed, resulting that the residual primary PEC portion reveals less agglomeration in system [2].

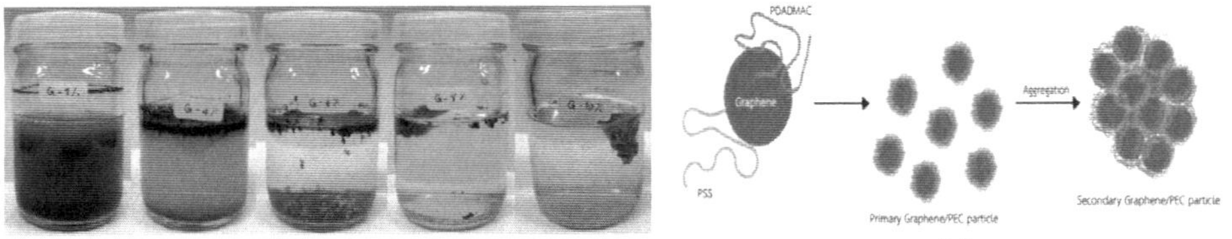

(a) (b)

Figure 2. (a) G/PDADMAC-PSS PECs at different NaCl concentrations (2, 4, 6, 8 and 10 %wt.) and (b) The mechanism of the formation of G/PECs

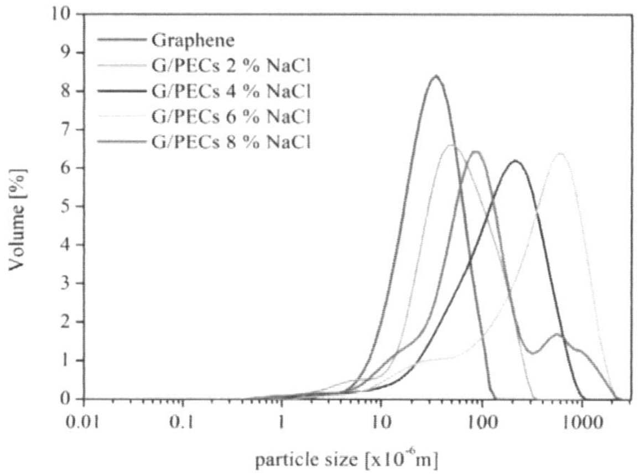

Figure 3. PSD curves of graphene and G/PDADMAC-PSS at different NaCl concentration

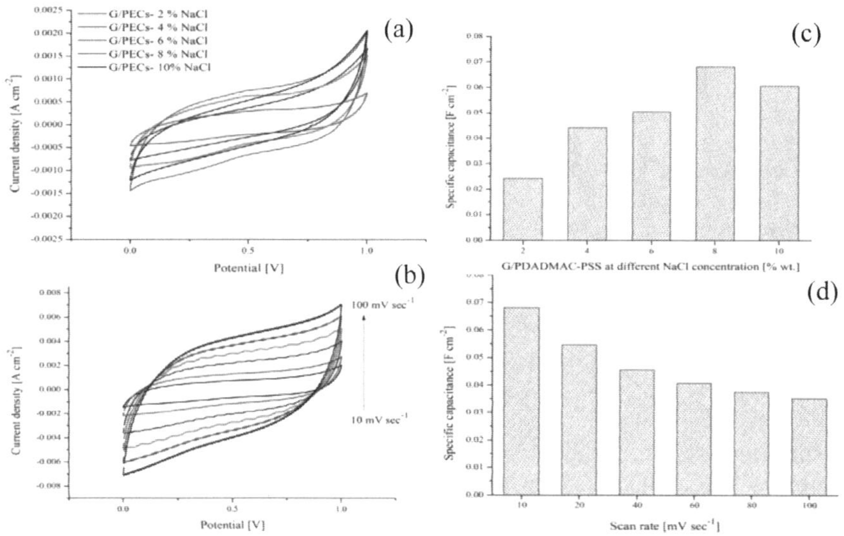

Figure 4. CV results in 1 M NaCl for (a) G/PDADMAC-PSS with different NaCl concentrations at 10 mV sec^{-1} (b) G/PDADMAC-PSS with 8 %wt. NaCl at different scan rate; The calculated specific capacitance values for (c) G/PDADMAC-PSS with different NaCl concentration at 10 mV sec^{-1} (d) G/PDADMAC-PSS with 8 %wt. NaCl at different scan rate

The comparative study of the CV plots is used to evaluate the electrochemical properties of the G/PDADMAC-PSS at different NaCl concentrations. Fig. 4a displays the CV curves for all those of G/PDADMAC-PSS on CFP electrodes at 10mV sec^{-1}, demonstrating the nearly idealized rectangular and symmetric shape implying a good electrical double layer capacitance behavior. The increasing of NaCl adding up to 8 %wt. affects the CV area by enhancing the electrochemical characteristics. This result might be arisen from the effect of more inserted Na^+ and Cl^- ion, inducing ion conductivity for capacitive properties. The slight decrease in CV area at 10 %wt. NaCl is due to the fact that the large aggregated G/PDADMAC-PSS is formed at high NaCl content, affecting to reduce in ion transportation.

The calculated the area specific capacitance values of G/PDADMAC-PSS at different NaCl concentration of 2, 4, 6, 8 and 10 %wt. are 24, 44, 50, 68 and 61 mF cm^{-1} at 10 mV sec^{-1}, respectively (see Fig. 4c). The capacitance preservations of G/PDADMAC-PSS at 8 %wt. NaCl is tested by increasing scan rates from 10 to 100 mVsec^{-1} as seen in Fig. 4b and 4d. The rectangularity of CV curve is deteriorated at very high scan rate, implying the development of electrochemical irreversibility. G/PDADMAC-PSS at 8%wt. NaCl electrode exhibits the capacitance preservations around 48 %.

4. Summary

G/PDADMAC-PSS PECs exhibit the size dependence at different NaCl concentrations by changing of the conformation of PEL molecules and the agglomeration process. This PDADMAC-PSS PECs at appropriate NaCl concentration can form the aggregated molecule which can apply on any shape surface. G/PDADMAC-PSS at adding 8 %wt. NaCl electrode can yield the highest specific capacitance of 68 mF cm^{-1} at 10 mV sec^{-1} with the capacitance preservations around 48 % which is good enough for practical capacitive electrode.

References

[1] V.S. Meka, Manprit K.G. Sing, M.R. Pichika, S.R. Nali, V.R.M. Kolapalli and P. Kesharwani: Drug Discov Today Vol. 22 (2017), p. 1697.Another reference
[2] V. Starchenko, M. Muller and N. Lebovka: J. Phys. Chem. B Vol. 116 (2012), p. 14961.
[3] K. Qingqing , W. John: J. materiomics Vol. 2 (2016), p. 37.
[4] L.L. Zhang and X.S. Zhao: Chem. Soc. Rev. Vol. 38 (2009), p. 2520.
[5] L. Xianbin, W. Na, W. Xiaoli and Z. Yuying: Nanomaterials and Nanotechnology Vol. 5 (2015), p. 1.
[6] C. Liu, Z. Yu, D. Neff, A. Zhamu and B. Z. Jang: Nano Lett. Vol. 10 (2010), p. 4863

IOP Conf. Series: Materials Science and Engineering **600** (2019) 012011 doi:10.1088/1757-899X/600/1/012011

Development and characterization of the mechanical properties of edible film from ginger starch, chitosan with glycerin as plasticizer to food packaging

Novia Sariningsih[1], Desi Suci Handayani[1], Triana Kusumaningsih[1*]

Chemistry Department, Faculty of Mathematic and Natural Science, Universitas Sebelas Maret, Indonesia

Corresponding e-mail: novia_sariningsih@student.uns.ac.id

Abstract. Packaging used edible film is one of the food preservation techniques. Many studies have proven that edible films can extend the shelf life and improve the quality of food products. Edible film from ginger starch and chitosan with glycerin as plasticizer has been successfully prepared and characterized. The purpose of this research was to produce and characterized of the mechanical properties of edible film from ginger starch, chitosan, and glycerin as plasticizer. Edible film was characterized by UTM, TGA, FTIR and Water Uptake Test. The results of tensile strength test of edible film is 0.67125 MPa and elongation up to 15.67%. The thermal degradation of material reached 176.9695°C, water uptake of edible film up to 2.5%, and FTIR data shows an interaction between ginger starch, chitosan, and glycerin. It concluded that edible film from ginger starch, chitosan, and glycerin as plasticizer has good mechanical properties and has potential as a packaging.

Keywords: edible film; chitosan; ginger starch;glycerine; mechanical properties.

1. Introduction

Packaging plays an important role to save the quality of food and to protect food from harmful contaminants, so it can improve food preservation. Plastics as food packaging has a negative impact on the environment [1]. Edible films can be used for increasing the food shelf-life because it have a good barriers against oxygen [2]. Edible films are commonly produced using carbohydrates, lipids, and proteins. Carbohydrates are often used for edible films because they are abundant in nature [3]. Edible films from carbohydrates have good characteristics and can protect packaged products from damage [4]. The elasticity of edible film from starch can be increased by using plasticizer such as glycerin [5].Chitosan has a good mechanical properties for edible film because it is able to form hydrogen bonds between chains with amylose and amylopectin in starch [6]. Chitosan is a natural preservative because it is antimicrobial, chitosan can inhibit bacterial growth [7]. Chitosan and oligomer are natural antimicrobials, anti-fungal[8],antitumor, function as hypocholesterolemic, insecticidal and fungicidal activity [9]. Ginger (*ZingiberofficinaleRosc.*) is a medicinal plant that contains bioactive components such as *gingerol,* (6)-*shogaol, diarilheptanoid* and *curcumin.* Ginger rhizome contains phenolic compounds and antioxidant activity better than *tocopherol*[10]. Ginger is often used as an ingredient in herbal medicines [11]. This study to show determine the effect of the composition of chitosan and ginger starch on the mechanical properties of edible film with glycerin as plasticizer so that it can protect well-packed food.

2. Experimental methods

2.1. Materials
The material used to produce edible film is commercial chitosan (from shrimp) with a deacetylation degree up to 87.5%, ginger starch, acetic acid (70% Dixi Merck), commercial glycerin, aquades.

2.2. Instruments
The instrument to preparation of edible film includes hotplate and magnetic stirrer C-MAG HS 7, beaker glass 500 mL Pyrex, pipette volume 10 mL Pyrex. Mechanical test of edible film consists of functional groups to find out the results of synthesis, tensile strength test, water uptake, and degradation thermal. The instrument includes FTIR Merck Shimadzu (Type IRprestige21), Testometric materials testing machines (Wintest^{TM} Analysis), and TGA-DTA Stanton Redcroft TG-500.

2.3. Edible film preparation
Ginger starch and chitosan are weighed with variations in the composition of ginger starch and chitosan (g/g), 1:9, 5:5, 9:1 with a total mass of ginger starch and chitosan is 2 gram. Ginger starch was dissolved using aquades and chitosan dissolved using acetic acid1% (Merck®, 100%) mixing for 30 minutes at 100°C. The two solution are mixed. The mixture was heating at 100°C and stirring for 25 minutes. After 25 minutes, the mixture is added 0.2 mL glycerin (10% from the solution) and stirred for 5 minutes or until thickened. The solution is taken 100 mL of the solution and place it on the mold, then leave it in the open space until the mold dries. The film is released from the mold and ready to be analyzed.

2.4. Characterization of edible film
Characterization of edible film includes tensile strength and elongation tests using Testometric Materials Testing Machines. The sample was clamp with a testing machine. The start button on the computer ispressed and then the tool will pull the sample at a speed of 5000 mm/minutes until the sample breaks. Tensile strength of edible film is calculated by the following equation.

$$\tau = \frac{\text{Fmax}}{\text{A}} \tag{1}$$

Based on equation above, τ indicated tensile strength (MPa), Fmax shows voltage (N), and A shows broad cross-section transverse (mm^2).
Measurement of elongation extention was carried with same methode for tensile strength, with following equation:

$$\text{Elongation (\%)}= \frac{\text{Strain isolate (mm)}}{\text{initial length (mm)}} \text{X } 100\% \tag{2}$$

While, for elasticity (modulus young) obtained from the comparison of tensile strength with elongation[12].
Functional group analysis with FTIR, sample placed into the set holder, then the appropriate spectrum is seeked. The result will be obtained by diffractogram relationship between wave number and intensity. The FTIR spectrum was recorded using a spectrophotometer at room temperature.
Water resistance test using the absorption of water, namely by measurement initial mass of the sample to be tested (Wo), then inserted into a container includes aquades during 10 minutes[13]. The sample is taken from a receptacle containing aquades and water contained in the plastic surface removed with paper tissue, then measured the water absorbed by the sample is calculated through equations:

$$\text{Water}\ (\%) = \frac{W - Wo}{Wo}\ X\ 100\% \tag{3}$$

Based on equation above, W is wet edible film mass, Wo is dry edible film mass.

TGA-DTA is a thermal analysis technique in a polymer. This technique is a dynamic method for recording sample mass under controlled or cooled conditions at a controlled rate as a function of time or temperature.Sample placed into the sample receptacle then heated, and seeking a spectrum that shows the reduced mass of the polymer changes in time or temperature due to the process of degradation or decomposition. Result will be obtained by diffractogram relationship between wave number and intensity.

3. Result and discussion

Testing the mechanical properties of edible films includes the strength test of tensile strength and elongation which from the test results shows the following data (Figure 1).

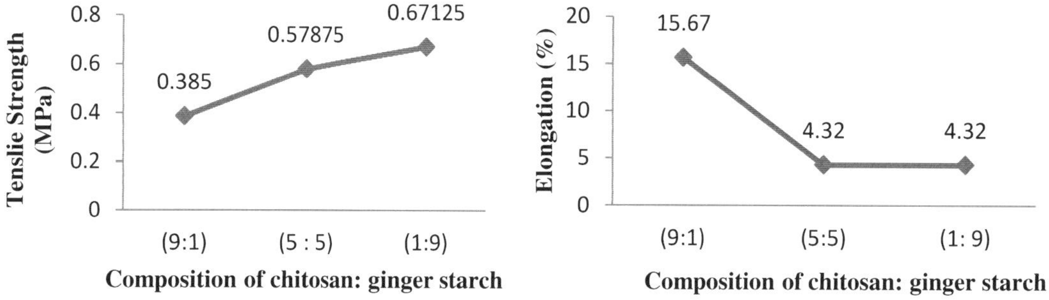

Figure 1. (a) Tensile Strength Test, (b) Elongation of Edible film

From these data indicate the maximum tensile strength is in the composition of chitosan: ginger starch 1:9 of 0.67125 MPa. This indicates that the more the amount of chitosan, the tensile strength of edible films decreases. This is probably due to the presence of chitosan which will interfere with the formation of the film matrix by the ginger starch polymer. This data is supported by elongation data from the following data (Figure 1b). where tensile strength is inversely proportional to elongation. An increase in elongation is affected by the presence of glycerin. Glycerin acts as a plasticizer that provides elastic properties to edible films. Glycerin has a small molecular weight so that it can enter into the intermolecular bond amylose with chitosan. The presence of plasticizers in the film can interrupt the formation of amylose with amylopectin branches, then reduce the interaction between amylose molecules and amylopectin, thereby increasing the flexibility of the film[14].

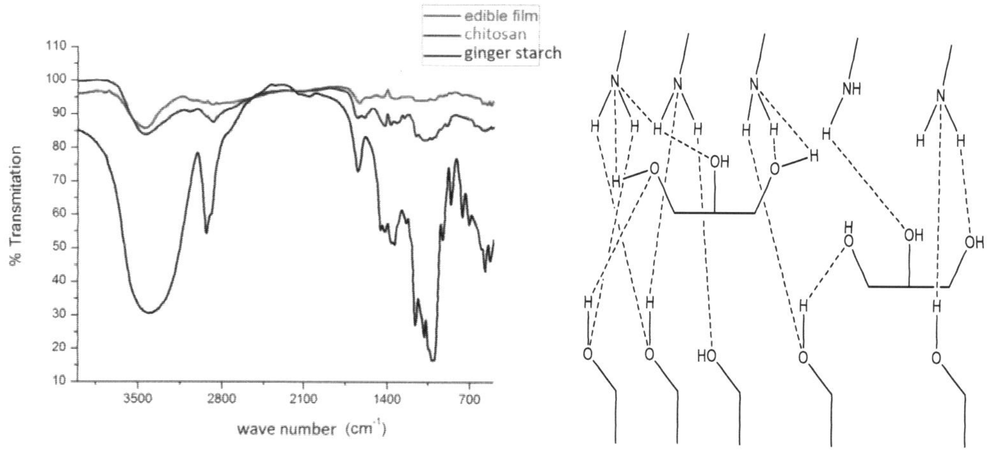

Figure 2. (a) FTIR spectra, (b) Interaction between ginger starch, chitosan and glycerin [15]

The interaction between ginger starch, chitosan and glycerin was shown through FTIR data in Figure 2. FTIR data shows the influence of the interaction between the –OH group on ginger starch, –NH group from chitosan and glycerin resulting in reduced free -OH group and amine group in the sample. Where uptake of the -OH group is around 3600 cm^{-1} 3200 cm^{-1}, while the-NH group (amine) is at absorption of 3500 cm^{-1}. From these interactions can be seen that the amine group from chitosan interacts with glycerin and ginger starch,resulting in reduced amine groups and free -OH groups, which supported by FTIR spectra, where in the edible spectra the resulting film has a lower -OH absorption and amine groups.

The water uptake test purpose to determine ability of edible film to absorb water.The smaller the ability to absorb water, the better the mechanical properties of the edible film. Edible films with well mechanical properties will able to protectwell-packaged products. Water uptake of edible film data is shown in figure 3a.

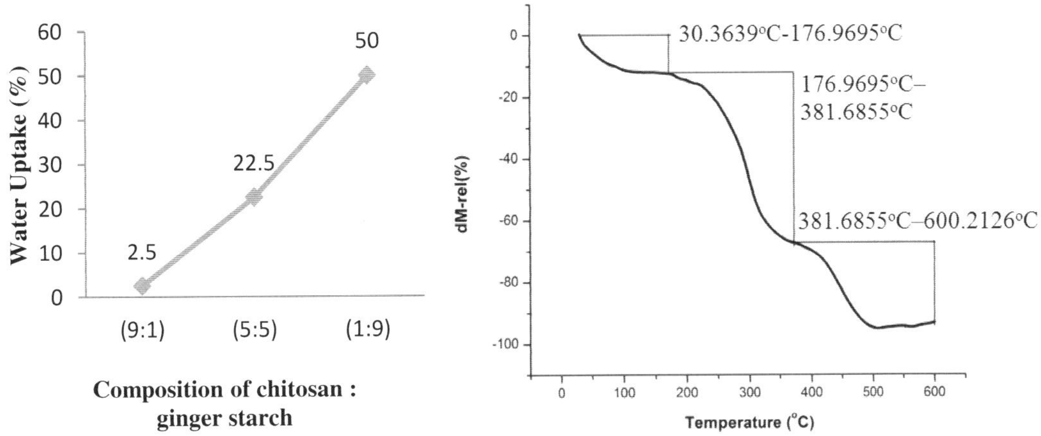

Figure 3.(a) ability to absorb water, (b) thermal degradation

From these data (3a) shows the greater the composition of chitosan, then the lower the percentage of water absorption from the edible film. Because the greater the amount of starch, the more the pores of the edible film will increase the ability to absorb water [16]. The higher the absorption of water, the mechanical properties of edible film are getting worse, because the absorption of water will interfere with molecular chain interactions, which are followed by increased diffusivity and able to absorb water vapor from the air[17]. TGA testing purpose to determine the degraded thermal temperature of the edible film, from the TGA data showing that the edible film have the degradation temperature at 176.9695°C. Thermal degradation of edible film data is shown in figure 3b.

Edible film of ginger starch and chitosan with glycerin as a plasticizer there are 3 stages of degradation which is at a temperature of 30.3639°C-176.9695°C is the evaporation of water, and then at temperature of 176.9695°C–381.6855°C is the degradation of ginger starch, and at temperature of 381.6855°C–600.2126°C is the degradation temperature of chitosan[18].

4. Conclusion

Based on the above data, it can be concluded that:

1. Edible film of ginger starch chitosan with glycerin as plasticizer has good mechanical properties. Maximum tensile strength is 0.67125 MPa, the best elongation up to 15.67%, the best water absorption ability is 2.5%, the degradation temperature is 176.9695°C, and the FTIR test results show an interaction between ginger starch, chitosan and glycerin.
2. The more the amount of chitosan added, the mechanical properties of edible film tend increase, because chitosan is able to close the surface pores of edible film and can increase the mechanical strength of the sample.

References

[1] Muscat, D., Adhikari, B., Adhikari, R., & Chaudhary, **2012**, Comparative study of film forming behaviour of low and high amylose starches using glycerol and xylitol as plasticizers, *J Food Eng* 109(2) 189-201.
[2] Cao, N., X. Yang, Y. Fu, **2009**, Effects of various plasticizers on mechanical and water vapor barrier properties of gelatin films, *Food Hydrocolloids* 23(3) 729-735.
[3] Kamper, S., O. Fennema, **1985**, Use of an edible film to maintain water vapor gradients in foods, *J. Food Sci* 50(2) 382-384.
[4] Guilbert, S., N. Gontard, and L.G.M. Gorris, **1996**, Prolongation of the shelf life perishable food products using biodegradable films and coatings. Lebensm. Wiss. *Technol.* 29: 10–17.
[5] Ekrami, M. and Emam-Djomeh, Z., **2014**, Water Vapor Permeability, Optical and Mechanical Properties of Salep-Based Edible Film, *Journal of Food Processing and Preservation*, *38*(4), pp.1812-1820.
[6] Chillo, S., Flores, M. Mastromatteo, A. Conte, L. Gerschenson, M. Del Nobile, **2008**, Influence of glycerol and chitosan on tapioca starch-based edible film properties, *J Food Eng* 88(2) 159-168.
[7] Agulló, E., Rondriguez, M.S., Ramos, V., dan Albertengo, L., **2003**, Present and Future role of chitin and chitosan in food. *Macromolecular Bioscience* 3(10): 521-530.
[8] Vilai, R., Nijarin, W., Nilada, K., dan Pachara, C., **2006**, Application of fungal chitosan for clarification of apple juice, *Journal Process Biochemistry.* 41:589-593.
[9] Entsar I.R., Mohamed, E.B., Tina, M.R., Christian, V.S., Monica, H., Walter, S., dan Guy, S., **2005**, Insecticidal and fungicidal activity of new synthesized chitosan derivatives. *Journal Pest Management Science* 61:951–960.
[10] Kikuzaki, N. Nakatani, **1993**, Antioxidant effects of some ginger constituents- *Journal of food science*, - Wiley Online Library.
[11] Thomson, M., Corbin, R., Leung, L., **2014**, Effects of Ginger for Nausea and Vomiting in Early Pregnancy: A Meta-Analysis. *Journal of the American Board of Famili Medicine.* 27(1), 115-122.
[12] Ban, W., Song, J., Argyropoulos, D. S. & Lucia L. A., **2005**, Improving the physical and chemical functionally of Starch – Derived Films with Biopolymers, *Journal of Applied Polymer Science* 100: 2542-2548.
[13] Sariningsih, N., Putra, Y.P., Pamungkas, W.P. and Kusumaningsih, T., **2018**, Edible Film from Polyblend of Ginger Starch, Chitosan, and Sorbitol as Plasticizer. In *IOP Conference Series: Materials Science and Engineering* 333(1), 012083.
[14] Zhang, P., Zhao, Y., Shi, Q., **2016**, Characterization of a novel edible film based on gum ghatti: effect of plasticizer type and concentration. *Carbohydr. Polym.* 153, 345-355.
[15] Agustin, Y.E., Karsono, S.P., **2016**, bioplastics synthesis from chitosan - starch kepok banana skin with the addition of additives. *Jurnal Teknik Kimia* 10(2), 40-48.
[16] Supeni, G., A. A. Cahyaningtyas., A. Fitrina, **2015**, Characterization of Physical and Mechanical Properties of Chitosan Added on Carrageenan and Modified Tapioca Edible Film. *J. Kimia Kemasan* 37(2), 103-110.
[17] Herliany, N. E., J. Santoso, E. Salamah, **2013**,Characteristics of Biofilms from Carrageenan. *Jurnal Akuatika* 4(1), 10-20.
[18] Martínez-Camacho, A.P., Cortez-Rocha, M.O., Ezquerra-Brauer, J.M., Graciano-Verdugo, A.Z., Rodriguez-Félix, F., Castillo-Ortega, M.M., Yépiz-Gómez, M.S. and Plascencia-Jatomea, M., **2010**, Chitosan composite films: Thermal, structural, mechanical and antifungal properties. *Carbohydrate Polymers*, *82*(2), pp.305-315.

IOP Conf. Series: Materials Science and Engineering **600** (2019) 012012 doi:10.1088/1757-899X/600/1/012012

Preparation of cobalt oxide on MXene/N, S-rGO surface for supercapacitors

G Threepiriyamongkol[1] and P Pattananuwat[2, 3, *]

[1]Department of Petrochemical and Polymer Science, Faculty of Science, Chulalongkorn University, Bangkok, Thailand
[2]Department of Materials Science, Faculty of Science, Chulalongkorn University, Bangkok, Thailand
[3]Research Unit of Advanced Materials for Energy Storage, Chulalongkorn University, Bangkok, Thailand

E-mail: *prasit.pat@chula.ac.th

Abstract. The electrodeposited cobalt oxide grown on MXene/N, S-rGO surface was successfully synthesised via chronoamperometry technique at -1.2 V. The effect of MXene/ N, S-rGO surface and electrodeposition time of cobalt oxide on the electrocapacitive behaviour was investigated. SEM-EDS evidenance confirmed that the highly dense growth of flake-like structure of cobalt oxide with interconnected pores was obtained on MXene/N, S-rGO surface. The synergistic effect of cobalt oxide grown on MXene/N, S-rGO surface can enhance the electrochemical properties with the highest specific capacitance of 608.3 F/g at 1 A/g (628.8 F/g at 10 mV/sec) with 35.7 Wh/kg and 180.6 kW/kg.

1. Introduction

Cobalt oxide is one of classic metal oxide applying for supercapacitors owning to its pseudocapacitor features. It can be easily synthesized in many techniques such as hydrothermal, sol-gel and electrodeposition technique, etc. Among of them, the electrodeposition technique is one of the most attractive attention methods because it can be deposited directly onto the substrate materials. This technique can be controlled the amount of cobalt oxide by the different deposited times. However, the limitation of electrodeposition technique is the lack of adhesion between the surface which depends on substrate types, causing the occurring of low yield of the deposited active materials. Thus, the modification surface of substrate is required to improve the surface activity. Several reports demonstrated that the using of materials coating, such as metal particle, carbon and so on, covering on the substrate can improve the surface adhesion. Recently, the discovery of the new electric storage materials, namely titanium carbide or MXene ($Ti_3C_2T_x$) possesses the 2D multilayer structure with a large surface area.

MXene has a theoretical hybrid capacitor attributing to redox reaction [1]. However, the restacking problem in its own morphological structure of MXene is reported, causing the loss of surface area for redox reaction [2]. Hence, the delamination of MXene layer is required to maintain 2D layer structure by adding organic solvent or small molecules such as dimethyl sulfoxide (DMSO) or carbon nanotubes (CNT) and graphene (rGO). The insertion of small molecules such rGO for intercalating between layer has been proved to exfoliate the self-restacking. In addition, the using of rGO doped nitrogen/sulfur atoms exhibited the beneficial features of more electron donor for enhancing the electrochemical performance [3]. Thus, in this work, the effect of MXene/N, S-rGO surface for the electrodeposited cobalt oxide growth is studied. The optimum condition for deposited time of cobalt oxide on MXene/N, S-rGO surface is also investigated to achieve the specific capacitance values.

2. Experimental section

2.1. Preparation of MXene/N, S-rGO coated materials on nickel foam

MXene was typically synthesized by etching of Al atom layers from the Ti_3AlC_2 precursors with HF, following by exfoliation process in dimethyl sulfoxide for 24 h. For graphene doped with nitrogen and sulfur atom, graphene oxide (GO) was synthesized by modified Hummer method [4]. GO was doped with nitrogen and sulfur atom, then partially reduced by reflux process in thiourea aqueous solution at equivalent weight, denoted as N, S-rGO. The obtained MXene and N, S-rGO were mixed at ratio of 95:5 in the aqueous solution under sonicate for 3 h. Subsequently, the mixture was dropped onto the nickel foam sheet and dry at 70 °C.

2.2. Preparation of cobalt oxide and cobalt oxide/MXene/N, S-rGO electrode samples

The cobalt oxide was electrochemically deposited from the solution of 0.05 M Co $(NO_3)_2$ $6H_2O$ via chronoamperometry technique at constant potential of -1.2 V. PGSTAT 12 equipped with a conventional three-electrode cell system consisting of MXene/N, S-rGO/nickel foam as working electrode, platinum as counter electrode and Ag/AgCl as reference electrode was used to employ the deposition process at various time of 60, 180 and 300 sec which are denoted as MxG-Co1, MxG-Co2 and MxG-Co3, respectively. The direct electrochemical deposition of cobalt oxide on the bare nickel foam substrate with similar process at 60, 180 and 300 sec, denoted as Co1, Co2 and Co3, respectively, was also done to compare the effect of MXene/N, S-rGO coated materials.

2.3. Characterization

The surface morphology and elemental analysis of samples were observed using field emission scanning electron microscopy (FE-SEM: JSM-7610F, JEOL) at an acceleration voltage of 15 keV. The samples were coated with 10 nm of gold before being observed.

2.4. Electrochemical measurements

The electrochemical properties of the as-prepared cobalt oxide/MXene/N, S-rGO electrode samples were examined in 6 KOH using three-electrode system configurating with platinum as electrode counter and Ag/AgCl as reference electrodes. PGSTAT 12 equipment was used for cyclic voltammetry (CV) measurements and galvanostatic charge/discharge (GCD). CV experiments were executed in the range of -0.3 to 0.45 V at different scan rates. The range of -0.2 to 0.40 V at different current densities was adjusted for measuring the GCD curves. The specific capacitance by CV was calculated according to the following equations [5].

$$C_{sp} = (\int IdV)/(2v \times m\Delta V) \tag{1}$$

Galvanostatic charge-discharge (GCD) measurement was carried out at various current densities. The specific capacitance by GCD was calculated according to the following equations [6].

$$C_{sp} = (I\Delta t)/(m\Delta V) \tag{2}$$

Where $\boldsymbol{C_{sp}}$ (F/g) is the specific capacitance, $\boldsymbol{v}$ (V/s) is the scan rate. $\boldsymbol{I}$ (A) is the discharge current. $\boldsymbol{\Delta t}$ (s) is the discharge time. $\boldsymbol{m}$ (g) is the mass of composite on substrate. And $\boldsymbol{\Delta V}$ (V) is the potential window without the hydrogen evolution.

3. Results and discussion

3.1. Morphology study

Figure 1a shows the SEM images of cobalt oxide on nickel foam at the deposition time of 180 sec, demonstrating the densely anisotropic growth of uniform flake structure derived from cobalt oxide. Obviously, the denser growth of flake-like structure of cobalt oxide with interconnected pores can be obtained from MXene/N, S-rGO surface for all the deposition times. The remarkable increase in dense flake-like structure with the increase deposition time is also presented for cobalt oxide deposited onto MXene/N, S-rGO surface. Some of aggregated cobalt oxide structure is found at the deposition time of 300 sec (Figure 1d). These evidences imply that the coated MXene/N, S-rGO possesses the beneficial feature for cobalt oxide deposition. Additional evidence from SEM-EDS reveals the existence of cobalt oxide, carbon and titanium on surface as seen in figure 1e, confirming coexistence of titanium carbide and cobalt oxide on surface electrode.

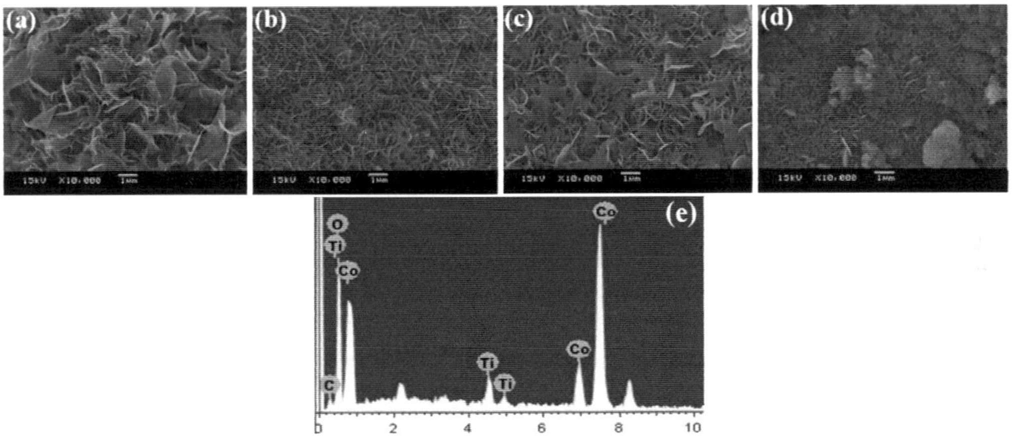

Figure 1. SEM images of (a) Co2, (b) MxG-Co1, (c) MxG-Co2, (d) MxG-Co3, (e) EDS analysis of MxG-Co2.

3.2. Electrochemical properties

CV curves in figure 2a show the electrochemical activity of the cobalt oxide electrodes at different deposition times. Figure 2b shows the CV curves of the cobalt oxide on MXene/N, S-rGO surface at different deposition times. As seen in figure 2a and 2b, the remarkable redox peaks of cobalt oxide electrodes and cobalt oxide on MXene/N, S-rGO obviously reveal the different faradic characteristics. All of cobalt oxide on MXene/N, S-rGO exhibits the higher in CV area than that of cobalt oxide on nickel foam, indicating the higher electrochemical activity. This evidence confirms the good synergistic effect between the cobalt oxide and titanium carbide causing in increasing in pseudocapacitance behaviour.

The calculated specific capacitance values by CV for cobalt oxide and cobalt oxide onto MXene/N, S-rGO at the different deposition times are presented in figure 2c. It is clear that all cobalt oxide deposited onto MXene/N, S-rGO samples show the highly efficient electrochemical activity surface than that of cobalt oxide deposited onto nickel foam. Among of MxG-Co series, MxG-Co2 can deliver the highest specific capacitive of 628.8 F/g. These results also correspond agreement with SEM results supporting that the high active surface of flake-like structure with interpore connection of MxG-Co2 can provide a facile shortcut path for charge transfer increasing a more efficient electrocapacitive properties. The slightly decrease of MxG-Co3 (489.7 F/g) is arisen from the aggregated cobalt oxide structure causing the decrease in charge transfer surface. The electrochemical performance of MxG-Co2 electrode at various scan rate is also evaluated as shown in figure 2d. A pair of redox peaks of samples was clearly observed for each curve and can be maintained, even at high scan rate. The good rate capability of MxG-Co2 electrode with capacitance preservations is around 55% (from 628.8 to 346.4 F/g).

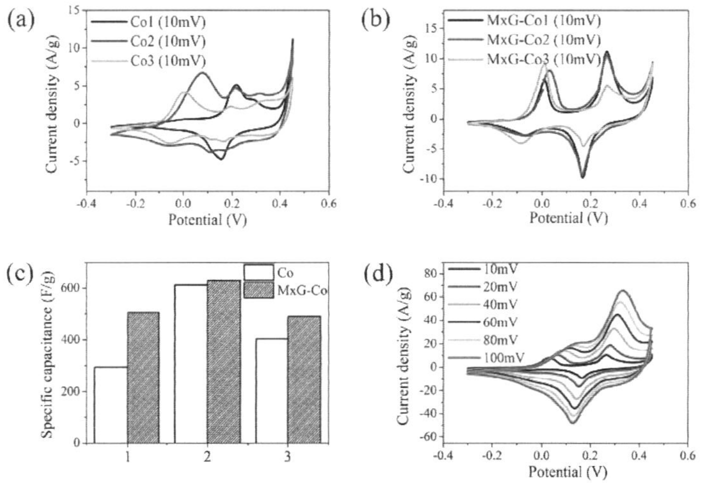

Figure 2. CV curves of (a) Co1, Co2 and Co3, (b) MxG-Co1, MxG-Co2 and MxG-Co3, (c) The calculated specific capacitance of samples, (d) CV curves of MxG-Co2 at various scan rates.

In addition, GCD curves in figure 3a show the charge-discharge profile of the cobalt oxide electrodes at different deposition time. Figure 3b shows the GCD curves of the cobalt oxide on MXene/N, S-rGO surface at different deposition time. All sample electrodes reveal the distinct pseudocapacitive characteristic with a presence of a large plateau in GCD curves. Among of electrode samples, MxG-Co2 reveals the longest plateau region for charge transfer reaction, indicating the highest capacitance. The specific capacitance (Cs) of cobalt oxide electrode samples calculated from discharge time at 1 A/g are displayed as figure 3c. The specific capacitance values at 1 A/g of Co1, Co2, Co3, MxG-Co1, MxG-Co2 and MxG-Co3 are 193.3, 365.83, 253.3 and 514.2, 608.3, 435.8 F/g, respectively. Figure 3d displays the good rate capability of MxG-Co2 with capacitance preservations of 57.5% (from 608.3 to 350 F/g) as the increasing of current density from 1 to 10 A/g, which is a good agreement with CV results.

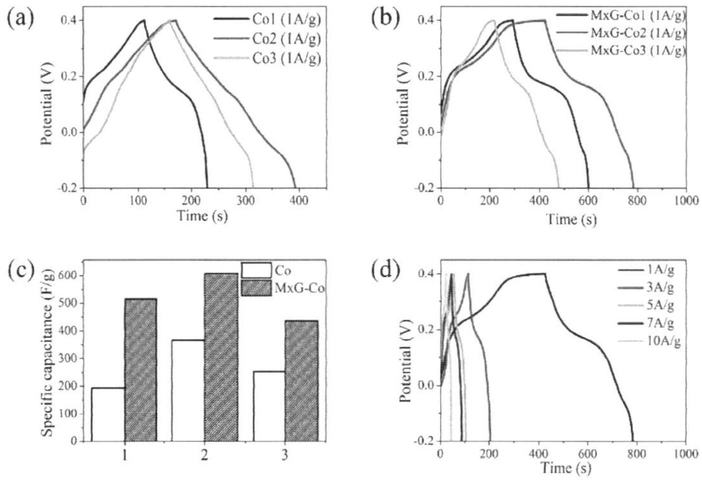

Figure 3. GCD curves of (a) Co1, Co2 and Co3 (b) MxG-Co1, MxG-Co2 and MxG-Co3, (c) The calculated specific capacitance of samples, (d) GCD curves of MxG-Co2 at various current densities.

4. Conclusions

The electrodeposited cobalt oxide grown on MXene/N, S-rGO surface has been developed to increase electrochemical supercapacitor performance by the beneficial features of the synergistic effects from pseudocapacitor mechanism between cobalt oxide and MXene/N, S-rGO surface. The high active surface of flake-like structure with interpore connection of cobalt oxide surface of MxG-Co2 can deliver the highest specific capacitance of 608.3 F/g at 1 A/g (628.8 F/g at 10 mV/s) with energy density of 35.7 Wh/kg and power density of 180.6 kW/kg.

References

[1] Sun S, Liao C, Hafez A M, Zhu H and Wu S 2018 Two-dimensional MXenes for energy storage, Chemical Engineering Journal Vol. **338**, pp 27-45.

[2] Li Z, Wang L, Sun D, Zhang Y, Liu B, Hu Q and Zhou A 2015 Synthesis and thermal stability of two-dimensional carbide MXene Ti_3C_2, Materials Science and Engineering: B Vol. **191**, pp 33-40.

[3] Wang T, Wang L X, Wu D L, Xia W and Jia D Z 2015 Interaction between Nitrogen and Sulfur in Co-Doped Graphene and Synergetic Effect in Supercapacitor, Scientific Reports Vol. **5**, pp 9591-9600.

[4] Hamlaoui Y, Tifouti L, Remazeilles C and Pedraza F 2010 Cathodic electrodeposition of cerium based oxides on carbon steel from concentrated cerium nitrate. Part II: Influence of electrodeposition parameters and of the addition of PEG, Materials Chemistry and Physics Vol. **120**, pp 172-180.

[5] Luo Y, Yang T, Zhao Q and Zhang M 2017 CeO_2/CNTs hybrid with high performance as electrode materials for supercapacitor, Journal of Alloys Compounds Vol. **729**, pp 64-70.

[6] Xiang C, Li M, Zhi M, Manivannan A and Wu N 2013 A reduced graphene oxide/Co_3O_4 composite for supercapacitor electrode, Journal of Power Sources Vol. **226**, pp 65-70.

IOP Conf. Series: Materials Science and Engineering **600** (2019) 012013 doi:10.1088/1757-899X/600/1/012013

Using borax as a cross-linking agent in cellulose-based hydrogels

Farin Phoothong[1], Anyaporn Boonmahitthisud[2,3*] and Supachok Tanpichai[4,†]

[1]Program of Petrochemical and Polymer Science, Faculty of Science, Chulalongkorn University, Bangkok, 10330, Thailand
[2]Department of Materials Science, Faculty of Science, Chulalongkorn University, Bangkok, 10330, Thailand
[3]Green Materials for Industrial Application Research Unit, Faculty of Science, Chulalongkorn University, Bangkok, 10330, Thailand
[4]Learning Institute, King Mongkut's University of Technology Thonburi, Bangkok, 10140, Thailand

*anyaporn.b@chula.ac.th and †supachok.tan@kmutt.ac.th

Abstract. In this research, the cellulose-based hydrogels were obtained from the water hyacinth fibers (WHFs) with aids of borax as a chemical cross-linker. The chemical treatment on WHFs, confirmed by Fourier transform infrared (FT-IR) analysis and thermogravimetric analysis (TGA), exhibited the reduction of hemicellulose and lignin. The treated cellulose fibers were dissolved in the mixed solution of NaOH/urea prior to mixing with borax at the cellulose/borax weight ratio of 1/0, 1/0.5 and 1/1 to prepare cellulose-based hydrogels. The effect of borax contents on the properties of cellulose-based hydrogels was investigated. The results showed that the swelling ratio and equilibrium water content of the obtained hydrogels were similar with increasing a borax content.

1. Introduction

Hydrogels, three-dimensional polymeric network materials containing large amounts of water, can be used in a wide range of applications such as agricultural, medical and waste-water treatment [1]. Generally, hydrogels can be derived from synthetic polymers such as poly(acrylic acid), poly(amido-amine), poly(ethylene glycol), and poly(vinyl alcohol) and from natural biomaterials for instance hyaluronate, alginate, starch, gelatin, chitin, chitosan, and cellulose [2]. In order to stabilize hydrogels without any dissolution, the physical or chemical cross-linking should be introduced. Physical cross-linking is usually obtained via a physical process such as chain entanglement, crystallization, and hydrogen bonding while chemical cross-linking is usually obtained via chemical bonding. The main disadvantage of the physical cross-linked hydrogels is the limited stability to long usage; therefore, chemical cross-linking would be suitable in this case [3]. Although synthetic polymer-based hydrogels have advantages over natural-based hydrogels such as adjustable mechanical properties and easy control of chemical compositions, they are non-renewable and non-biocompatible compared with natural-based hydrogels [4]. Recently, cellulose, a renewable resource found in plants such as straw, corncob, bagasse, and water hyacinth, has been widely used to prepare hydrogels due to its favourable properties including hydrophilicity, biodegradability, biocompatibility, nontoxicity, and low cost [2].

In this research, borax was introduced as a cross-linker in order to enhance stability and properties of the hydrogels prepared from water hyacinth fibers (WHFs), and effect of the borax contents on hydrogel properties was investigated.

2. Experimental

2.1. Materials

WHFs were obtained from Phra Nakhon Si Ayutthaya province, Thailand. Ethanol and toluene were purchased from CT Chemical. Hydrochloric acid (HCl) was purchased from JT Baker Chemicals, and sodium hydroxide (NaOH) and sodium hypochlorite (NaClO) were purchased from Loba Chemie. Urea and borax were purchased from Ajax Finechem and Quality Reagent Chemical, respectively. All materials were used as received without further purification.

2.2. Preparation of cellulose hydrogels

Prior to preparation of cellulose hydrogels, WHFs were first treated by chemical method. First, WHFs were dewaxed using a mixture of toluene-ethanol solution (2:1, v/v) at 73 °C for 3 hours. The obtained dewaxed fibers were filtered, washed with deionized (DI) water and then oven-dried at 60 °C for 24 hours. 10 g of these dewaxed fibers were added to 500 ml of 3 vol% of NaClO aqueous solution for 2 hours at 80 °C. After that, alkaline treatment was performed by 1 wt% of NaOH aqueous solution for 2 hours at 60 °C. Next, the second bleaching process was performed using 3 vol% of NaClO aqueous solution for 3 hours at 75 °C. Later, hydrolysis was performed by 5 vol% of HCl solution for 6 hours at 60 °C. The treated cellulose fibers were washed with excess distilled water to neutral and then freeze-dried for 60 hours. The cellulose solution was obtained by dissolving 3 g of the treated cellulose fibers in 100 ml of the mixed solution of 7 wt% of NaOH and 12 wt% of urea [5]. The solution with the cellulose concentration 3 wt% was mixed with borax as a cross-linking agent with three different cellulose/borax weight ratios (1/0, 1/0.5 and 1/1) using a magnetic stirrer at room temperature. Later, the 10 ml of the obtained solution was poured into a Petri dish, and allowed to dry at room temperature for 5 days to form hydrogels.

2.3. Characterization

Functional group analysis of the raw WHFs, treated WHFs and cellulose hydrogels was carried out using Fourier transform infrared spectroscopy (FT-IR) (Nicolet 6700, Thermo Fisher Scientific) over a frequency range of 450 - 4000 cm^{-1}.

Thermal stability of the raw WHFs and treated WHFs was studied by the weight loss from thermal degradation using thermogravimetric analysis (TGA) (TGA/SDTA 851^e, Mettler Toledo). The samples were scanned from 30 to 700 °C at a heating rate of 20 °C min^{-1} under the nitrogen atmosphere with a flow rate of 40 ml min^{-1}.

The swelling ratio of the hydrogels was investigated at room temperature in DI water by using gravimetric method. The dried hydrogels were swollen in DI water for 24 hours to achieve equilibrium swelling. The weights of the dried hydrogels (W_d) and the swollen hydrogels (W_s) were determined to calculate the swelling ratio.

$$\text{swelling ratio (\%)} = \left(\frac{W_s - W_d}{W_d}\right) \times 100 \quad (1)$$

Equilibrium water content (EWC) is the amount of absorbed water in swollen hydrogels calculated by the following equation.

$$\text{EWC (\%)} = \left(\frac{W_s - W_d}{W_s}\right) \times 100 \quad (2)$$

3. Results and discussion

Figure 1(a) presents the color change of the WHFs after a series of the chemical treatment from brownish to white fibers. This was due to the reduction of lignin caused by alkaline treatment and bleaching process. Moreover, the fiber yield after these treatments was 42.7 %. Figure 1(b) shows FT-IR spectra of the raw and treated fibers. Both of the raw and treated WHFs reveal the peaks at 3400 and 2923 cm^{-1}, associated with O-H stretching vibration and C-H stretching vibration, respectively. In addition, the peaks located at 1054 and 1101 cm^{-1}, assigned to C-O primary alcohol and C-O secondary alcohol of cellulose are observed from both fibers. Moreover, the reduction in the intensity of the peaks at 1727 and 1519 cm^{-1}, assigned to C=O and C=C aromatic of hemicellulose [6] and at 1245 cm^{-1}, contributed to C-O stretching vibration of hemicellulose and aryl alkyl ether groups in lignin [7] was found from the treated WHFs, compared with the raw fibers. This confirmed the decrease in hemicellulose and lignin contents after the chemical treatments.

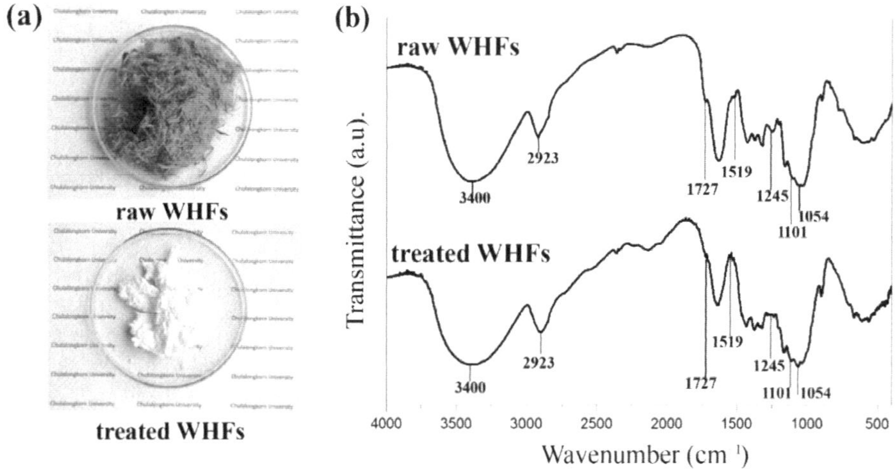

Figure 1. (a) Appearance and (b) FT-IR spectra of the raw and treated WHFs.

The Thermogravimetric (TG) curves of the raw and treated WHFs are shown in Figure 2(a). Moreover, Table 1 summarizers the thermal degradation of both fiber materials including onset temperature (T_{onset}), endset temperature (T_{endset}), maximum degradation temperature (T_{max}) of each stage and % char at 700 °C. The T_{onset} and T_{endset} thermal degradation temperature of the treated WHFs were 326 and 375 °C while the raw fibers presented the T_{onset} and T_{endset} of 269 and 349 °C, respectively. It can be seen that the thermal degradation temperature of the treated WHFs was higher than that of the raw WHFs, indicating the improvement of the thermal stability due to the removal of the hemicellulose and lignin. Generally, natural fibers present four thermal degradation regions: moisture evaporation, decomposition of hemicellulose and cellulose, degradation of lignin, and formation of pyrolytic char. However, these steps in a TGA curve were overlapped because of the thermal degradation temperature ranges of hemicellulose, cellulose, and lignin. Thus, a derivative thermogram (DTG) curve was used to present the thermal degradation regions, as shown in Figure 2(b). The initial degradation was observed at the range of 41 – 121 °C with $T_{max,1}$ of 60 °C, attributed to moisture evaporation. The second step was observed at the range of 195 – 387 °C with $T_{max,2}$ of 319 °C, attributed to the degradation of hemicellulose and cellulose while the third step was the degradation of lignin occurring between 417 – 511 °C with $T_{max,3}$ of 460 °C. The final step of the degradation happened at the range of 577 – 639 °C with $T_{max,4}$ of 617 °C, assigned to the formation of carbon [8]. On the other hand, the degradation of treated WHFs occurred only two steps. In the first step, moisture was evaporated at the range of 39 – 93 °C with $T_{max,1}$ of 48 °C and the second step was observed between 220 – 402 °C with $T_{max,2}$ of 359 °C. The third and fourth steps of the treated WHFs did not find due to a small amount of lignin left in the fibers. Moreover, the raw and treated WHFs presented the residue content at 700 °C of 17.2 and 9.1 wt%, respectively. The higher char content obtained from the raw fibers was due to the formation of pyrolytic char from lignin, hemicellulose, and some inorganic.

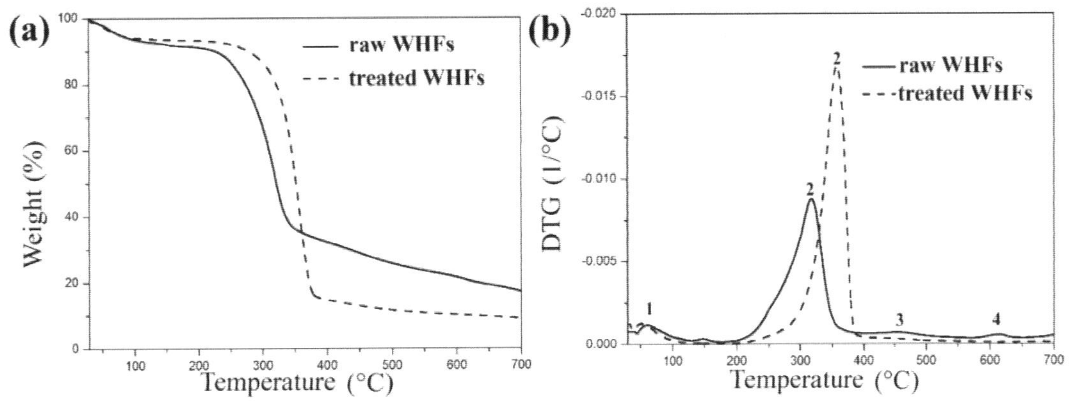

Figure 2. (a) TG and (b) DTG thermograms of the raw and treated WHFs.

Table 1. TG and DTG data obtained from the raw and treated WHFs.

Samples	T_{onset} (°C)	T_{endset} (°C)	$T_{max,1}$ (°C)	$T_{max,2}$ (°C)	$T_{max,3}$ (°C)	$T_{max,4}$ (°C)	% char at 700 °C
raw WHFs	269	349	60	319	460	617	17.2
treated WHFs	326	375	48	359	-	-	9.1

Figure 3(a) shows the appearance of the cross-linked cellulose hydrogels at the ratios of cellulose/borax of 1/0, 1/0.5 and 1/1. The cellulose hydrogels were transparent, indicating the good dissolution of cellulose in the NaOH/urea aqueous solution. However, the color of the cellulose hydrogels was light yellow. This might be due to a small amount of lignin remained in the treated WHFs. FT-IR spectra of the cross-linked cellulose hydrogels are presented in Figure 3(b). The cellulose hydrogels exhibited the characteristic absorption peaks at 3400, 2923, 1054, and 1101 cm^{-1}, associated with O-H stretching vibration, C-H stretching vibration, C-O primary alcohol, and C-O secondary alcohol, respectively. In addition, the reduction in the peak intensity at 3400 cm^{-1} with increasing a borax content was observed due to the decrease of the hydroxyl groups of cellulose. When borax was introduced into the cellulose hydrogel, the interaction between borax and hydroxyl groups was formed [9]. Moreover, the appearance of the trivial peaks at 1334 and 1278 cm^{-1}, assigned with borate (asymmetric stretching relaxation of B-O-C) [10] and vibrations of B-O-C, confirmed the cross-linking reaction between cellulose and borax (Figure 3(c)). Notably, the cross-linked cellulose hydrogels exhibited lower intensity of the peaks at 1334 and 1278 cm^{-1} were compared with others research, due to the low content of borax used in this study [9] [10].

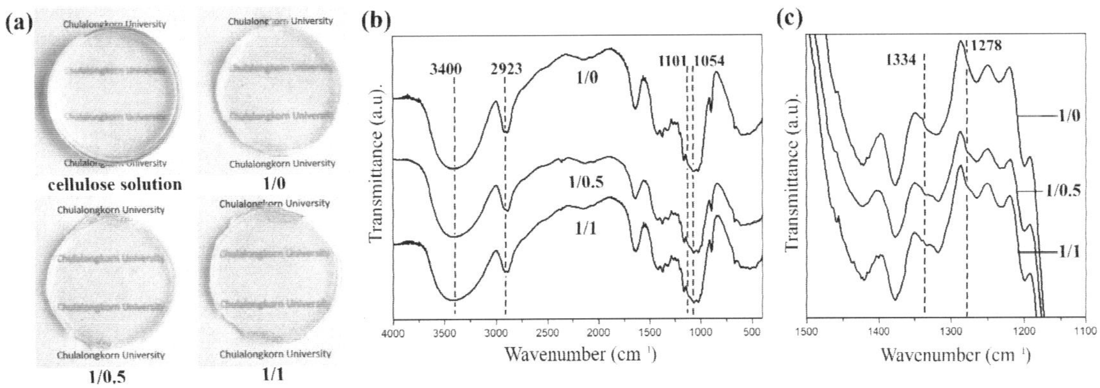

Figure 3. (a) Appearance of cellulose solution and cellulose hydrogels, (b) FT-IR spectra of the cellulose hydrogels, and (c) FT-IR spectra of cellulose hydrogels at 1500–1100 cm^{-1}.

The swelling ratio and equilibrium water content (EWC) of the varied cellulose/borax weight ratios of the cross-linked cellulose hydrogels are shown in Table 2. It was observed that the values of the swelling ratio of the cellulose hydrogels did not significantly change when the weight ratio of the borax increased. The swelling ratio of the cross-linked cellulose hydrogels with the ratio of cellulose/borax of 1/0, 1/0.5, and 1/1 was 75.1 ± 1.3, 79.2 ± 1.6, and 83.2 ± 2.6, respectively. The EWC of the cross-linked cellulose hydrogels with the ratio of cellulose/borax of 1/0, 1/0.5, and 1/1 was 89.9 ± 0.9, 90.7 ± 0.4, and 91.0 ± 0.5, respectively. The increase of the borax content did not affect swelling ratio and EWC at the 95% confidence interval. The similar values of these swelling ratio and EWC might be because of the low content of borax and the high hydrophilicity of cellulose interacted with water. The study of the effects of borax on properties of the prepared hydrogels is our future work to deeper understand the mechanism and the role of borax in the cellulose hydrogels.

Table 2. Swelling ratio and equilibrium water content of the cellulose hydrogels

Weight ratios of cellulose/borax	Swelling ratio (%)	EWC (%)
1/0	75.1 ± 1.3[a]	89.9 ± 0.9[b]
1/0.5	79.2 ± 1.6[a]	90.7 ± 0.4[b]
1/1	83.2 ± 2.6[a]	91.0 ± 0.5[b]

[a,b] **The same alphabet means no different statistical meaningful at a 95% confidence interval.**

4. Summary

The cross-linked cellulose-based hydrogels derived from WHFs were successfully prepared by chemical cross-linking with borax. The WHFs were firstly treated by the chemical treatment to remove impurities, hemicellulose and lignin. The yield of the treated WHFs was about 42.7%. The FT-IR spectra and TGA could confirm the reduction of hemicellulose and lignin contents in the treated WHFs. The treated cellulose was used to prepare the cellulose hydrogels with various amounts of borax. The appearance of the functional groups between borax and hydroxyl groups of cellulose observed from FT-IR results could confirm the cross-linking in the hydrogels. However, the addition of borax did not affect the swelling content and EWC values of the hydrogels.

Acknowledgments

The authors are grateful to Chulalongkorn University and Faculty of Science, Chulalongkorn University for financial supports.

References

[1] M. R. Guilherme, A. V. Reis, A. T. Paulino, T. A. Moia, L. H. Mattoso, and E. B. Tambourgi: Journal of Applied Polymer Science. Vol. 117 (2010), p. 3146-3154
[2] C. Chang and L. Zhang: Carbohydrate Polymers. Vol. 84 (2011), p. 40-53
[3] V. K. Thakur and M. K.Thakur: Handbook of Polymers for Pharmaceutical Technologies, Processing and Applications. Vol. 2 (2015), p. 95-120
[4] J. Zhu: Biomaterials. Vol. 31 (2010), p. 4639-4656
[5] H. Qi, C. Chang, and L. Zhang: Cellulose. Vol. 15 (2008), p. 779-787
[6] M. Thiripura and A. Ramesh: Carbohydrate Polymers. Vol. 87 (2012), p. 1701-1705
[7] Z. Wang, Z. Yao, J. Zhou, M. He, Q. Jiang, S. Li, Y. Ma, M. Liu, and S. Luo: Biological Macromolecules. Vol. 129 (2018), p. 1081-1089
[8] B. T. N. Thi, L. K. Ong, D. T. N. Thi, and Y. H. Ju: Journal of the Taiwan Institute of Chemical Engineers. Vol. 71 (2017), p. 55-61
[9] H. Bian, L. Jiao, R. Wang, X. Wang, W. Zhu, and H. Dai: European Polymer Journal. Vol. 107 (2018), p. 267-274
[10] M. Lim, H. Kwon, D. Kim, J. Seo, H. Han, and S. B. Khan: Progress in Organic Coatings. Vol. 85 (2015), p. 68-75

IOP Conf. Series: Materials Science and Engineering **600** (2019) 012014 doi:10.1088/1757-899X/600/1/012014

Phase Evolution - Property Relationships of PIMed 5 - 40 Vol.% SiCp - Reinforced Aluminium Composite

Tapany Patcharawit[1,a], Arada Ngeekoh[1,a], and Nutthita Chuankrerkkul[2,b]

[1]School of Metallurgical Engineering, Suranaree University of Technology, 111 University Avenue, Suranaree Subdistrict, Meung, Nakhon Ratchasima, 30000, Thailand

[2]The Metallurgy and Materials Science Research Institute, Chulalongkorn University, Phayatai Rd., Pathumwan, Bangkok, 10330, Thailand

[a]tapany@sut.ac.th; [b]Nutthita.C@chula.ac.th

Abstract. Microstructure-mechanical property relationships of powder injection molded aluminium composite reinforced with 5 - 40 vol.% silicon carbide particulate (SiC_p) has been investigated via thermal and phase analyses along with density and hardness testing. The samples were injection molded and sintered at 645 - 800 °C in N_2 atmosphere. Thermal analysis showed endothermic reactions around 632 - 643 °C, related to liquid phase formation. Exothermic reactions about 458 - 461 °C and weight gains around 471 - 479 °C were plausibly associated with AlN formation, which facilitated liquid phase sintering. The composite microstructure exhibited close interfaces between SiC_p - Al and Al - Al for 5 - 20 vol.% SiC_p addition. Higher SiC_p addition however led to SiC_p clustering, creating large pores and unreacted SiC_p - SiC_p interface, which prohibited both liquid flow and rearrangement of the solid grains and pore filling during liquid phase sintering. Optimum SiC_p addition was achieved at 15 vol.%, giving the maximum bulk density at 2.68 g.cm^{-3} or at 94.2 % theoretical density. The maximum macro Vickers hardness values at 138.9 and 166.6 H_v were obtained in the as - sintered and age - hardened conditions respectively.

Keywords: thermal analysis; phase evolution; aluminium composite; powder injection molding, properties

1. Introduction

Composite materials have been of great interest for recent years for many engineering applications such as electrical, biomedical and automotive. Fabrication techniques in either solid, semi-solid or liquid routes yield different interfaces among the reinforcements and the matrices, which finally determine the composite properties. Powder injection molding or PIM is recognized as low cost and mass productive for complicate-small components. The use of PIM for composite fabrication has gained attention due to feasibility in powder formulation and process parameters [1-4]. However to achieve optimum properties, care has to be taken into account along the course of feedstock preparation, injection molding, debinding and sintering. In liquid phase sintering full densification is desirable, where situations such as wetting of liquid on solid grains and solubility of solid in liquid are required [5]. This research investigated phase evolution during liquid phase sintering of 5 - 40 vol.%

SiC_p - reinforced aluminium composite in N_2 atmosphere, associated with its density and hardness in order to identify the optimum SiC_p addition suitable for the PIM process.

2. Experimental procedure

The aluminium composite of varying SiC_p contents at 5, 10, 15, 20, 30 and 40 vol.% was fabricated by powder injection molding using 73 µm Al - Cu powder of 4.77 wt.% Cu, 1.11 wt.% Mg, 1.01 wt.% Si, 0.12 wt.% Fe composition and SiC_p of 20 µm size. Feedstocks prepared at 55% solid loading were subjected to Simultaneous Thermal Analysis (STA) using a NETZSCH STA 449 F3 Jupiter operating at a 40 K/min heating rate up to 900 °C in a N_2 feeding at 50 ml/min for the determination of liquid phase formation. Injection molding was operated at 170 °C to produce green samples. After solvent debinding, sintering was carried out at 645 - 800 °C using a 1 L/min N_2 flow rate. The sintered samples were subjected to age-hardening at 150 °C/6 h. Phase analysis was carried out using a JEOL - JSM 6061LV SEM + EDS and a D8 XRD analyzer.

3. Results

3.1. Simultaneous thermal analysis

STA result including Differential Scanning Calorimetry (DSC) and Thermogravimetric (TG) of feedstock prepared at 40 vol.% SiC_p addition is demonstrated in Fig. 1, while Figs. 2 a) and b) demonstrate the onsets of endo/exothermic reactions (DSC1 - DSC5) and the onsets of weight loss (TG1 - TG5). Over the investigated temperature range up to 900 °C, thermal reactions and mass losses help to preliminarily determine suitable PIM and liquid phase sintering parameters, which are; i) the molding temperature, ii) the disposal of the multi-component binder and iii) the onset of liquid phase sintering for each formulation of the feedstock as follows:

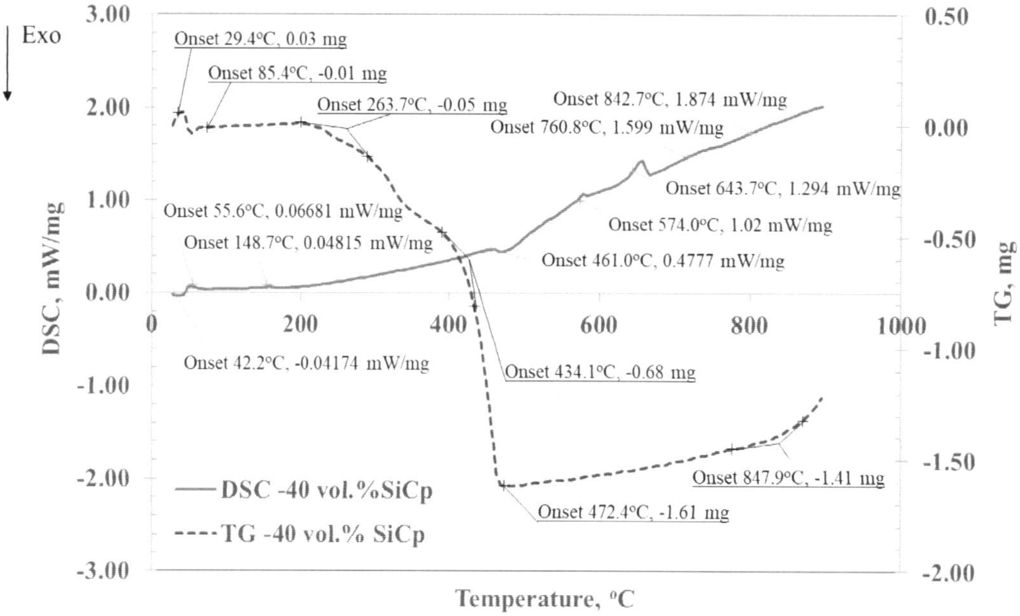

Figure 1. DSC and TG analyses of 40 vol.% SiC_p feedstock up to 900 °C in N_2 atmosphere.

Firstly, DSC data shows the onset temperatures at 55.6 (DSC1) and 148.7 °C (DSC2), related to the melting temperatures of the multi-component binder. Molding temperature is then set at 170 °C to allow efficient feedstock flow into mold cavity and enable green scrap reuse without burning.

Secondly, the TG graph reveals the onset of a small mass loss at 84.5 °C (TG1) due to vaporization of adsorbed water. Then significant mass losses at 263.7 °C and finished at 472.4 °C (TG2 - TG3), agreed with the vaporization of the binders. Therefore, soaking at 500 °C/0.5 h. was to assure the final disposal of the skeleton binder before reaching liquid phase sintering.
Thirdly as the temperature raised, endothermic reactions around 632 - 643 °C (DSC6), suggests the liquid phase formation. However, the actual sintering temperature had to raise up to 645 °C to 660 °C for aluminium composite reinforced with 5 - 20 vol.% SiC_p addition in order to get effective sintering. In the case of 30 - 40 vol.% SiC_p addition, the sintering temperature was 780 - 800 °C, in which refractoriness and agglomeration of the SiC_p became prevalent. This will be later discussed.

3.2. XRD analysis
Comparative XRD results of as - sintered and age - hardened samples are shown in Figs. 3 a) and b) respectively for 40 vol.% SiC_p reinforced aluminium composite. Both conditions share similar XRD peaks comprising the main spectra for Al and SiC as the matrix and reinforcement. AlN spectra were observed as the result of N_2 sintering gas. Small spectra of Al_2Cu and Mg_2Si were detected according to the aluminium alloy powder composition.

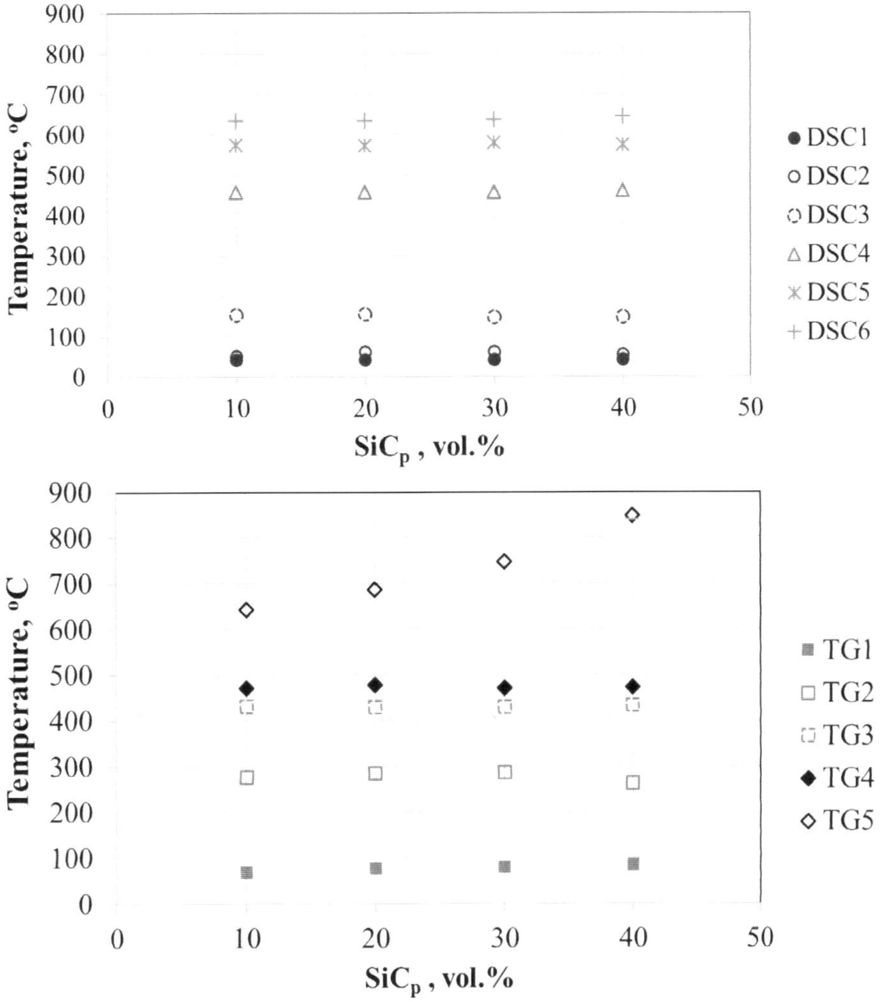

Figure 2. Onset temperatures corresponding to a) thermal reactions and b) weight losses of feedstocks at varying SiC_p additions.

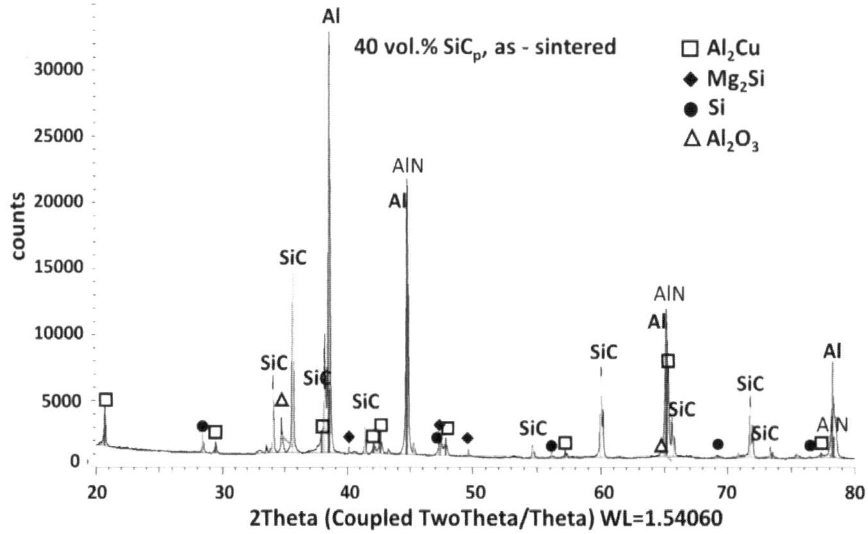

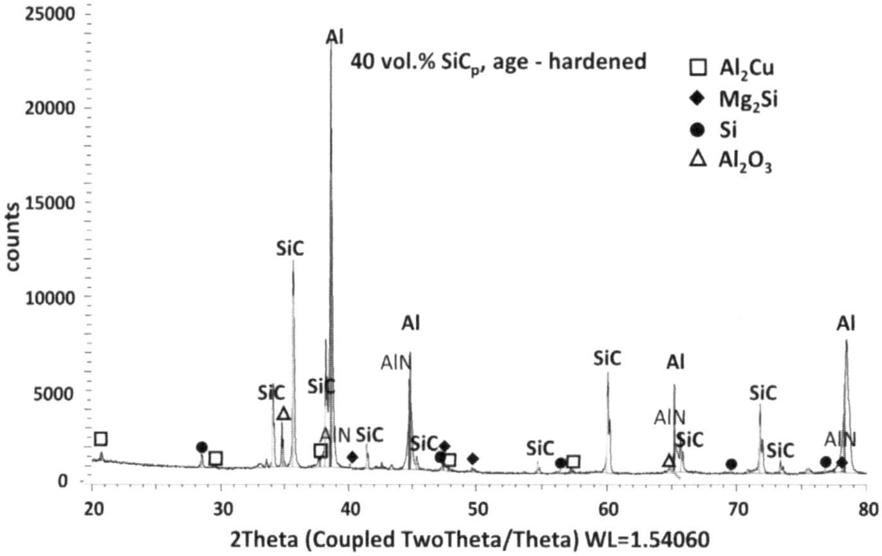

Figure 3. XRD results of 40 vol.% SiC_p reinforced aluminium composite in as - sintered (top) and age - hardened (bottom) conditions.

3.3. SEM and EDS analysis

Microstructures of aluminium, SiC_p and feedstock are illustrated in Figs. 4 a) - d). It is noticed that the aluminium powder surface has become rougher after injection molding. It is possibly that the abrasive nature of angular shaped SiC_p has scratched on the aluminium powder surfaces during feedstock preparation via ball-mill mixing, feedstock preparation, and powder injection molding. The disrupted oxide surface has later shown to facilitate magnesium migration and aluminium oxide consumption, favourable for liquid phase sintering. Discussion will be later introduced in this regard. In addition, poor particle packing is obtained at high SiC_p content (30 - 40 vol.%) due to high surface energy of SiC_p, leading to SiC_p clustering along with large inter - particle porosity as shown in Figs. 5 e) and f), which is difficult to be rid of, even after liquid phase sintering. Macroscopic pores were found along large SiC_p clusters at high SiC_p content.

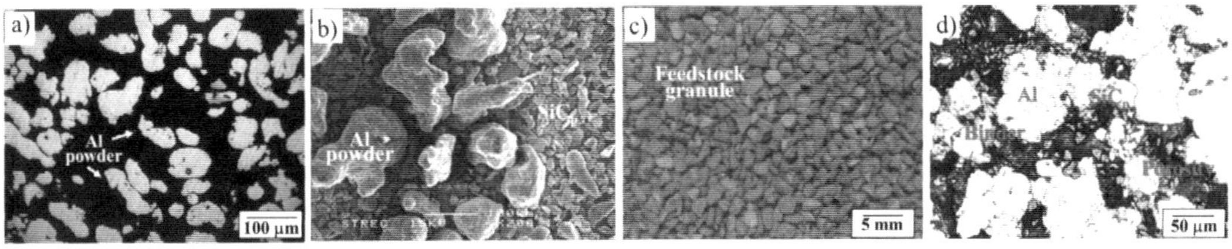

Figure 4. Powder morphology of a) aluminium alloy powder cross-sections, b) aluminium alloy powder and SiC_p, c) feedstock and d) microstructure of feedstock at 10 vol.% SiC_p addition.

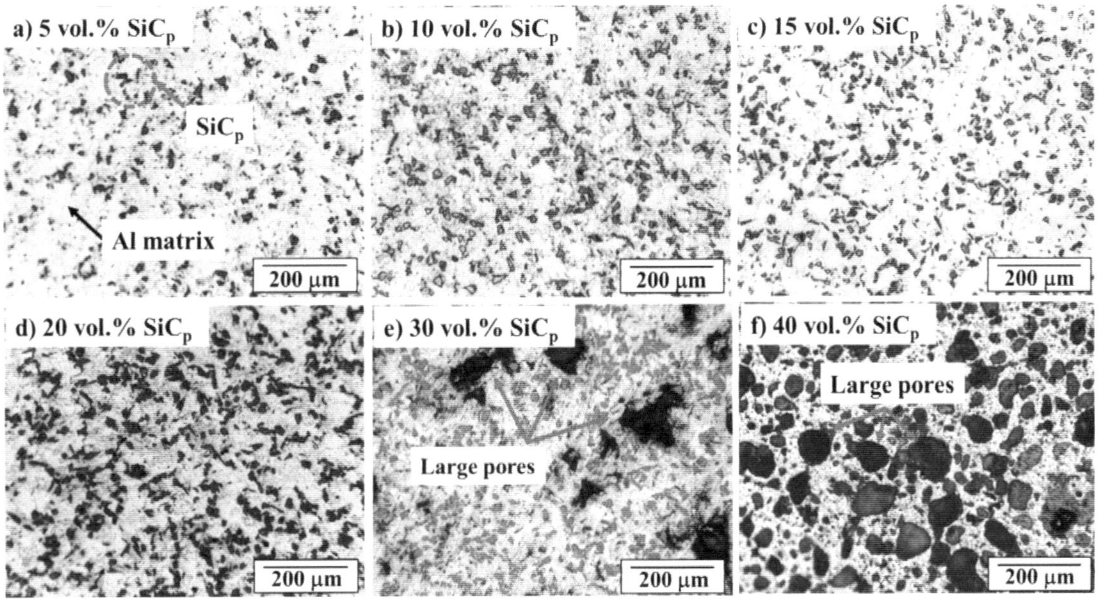

Figure 5. Optical micrographs of 5-40 vol.% SiC_p reinforced aluminium composite.

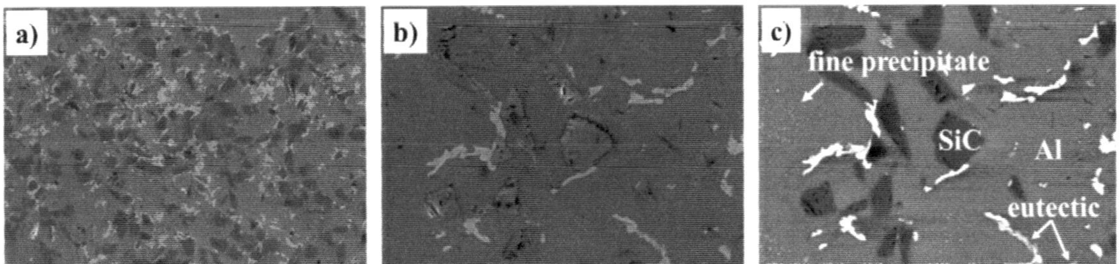

Figure 6. SEM micrographs of age - hardened SiC_p - aluminium composite a) in as secondary electron mode, b) at higher magnification and c) in back - scattered electron mode.

SEM investigation reveals uniform distribution of SiC_p and good interfaces between SiC_p - Al and Al - Al for the composite containing low SiC_p additions of 5 - 15 vol.%, as shown in Figs. 6 a) - c). Back scattered electron image displays contrast among SiC_p (dark), aluminium alloy matrix (grey) and eutectic phase (bright). Such interfaces among these phases confirms liquid-phase sintering taking place. Fine precipitates can be differentiated from the aluminium matrix. EDS point analysis confirms compositions of aluminium matrix (spectrum 1), eutectic phase (spectrum 2), SiC_p (spectrum 3) and precipitate phase of possibly Al_2Cu (spectrum 4), as shown in Figs. 7 a) - e) for 40 vol.% SiC_p reinforced aluminium composite in the age - hardened condition.

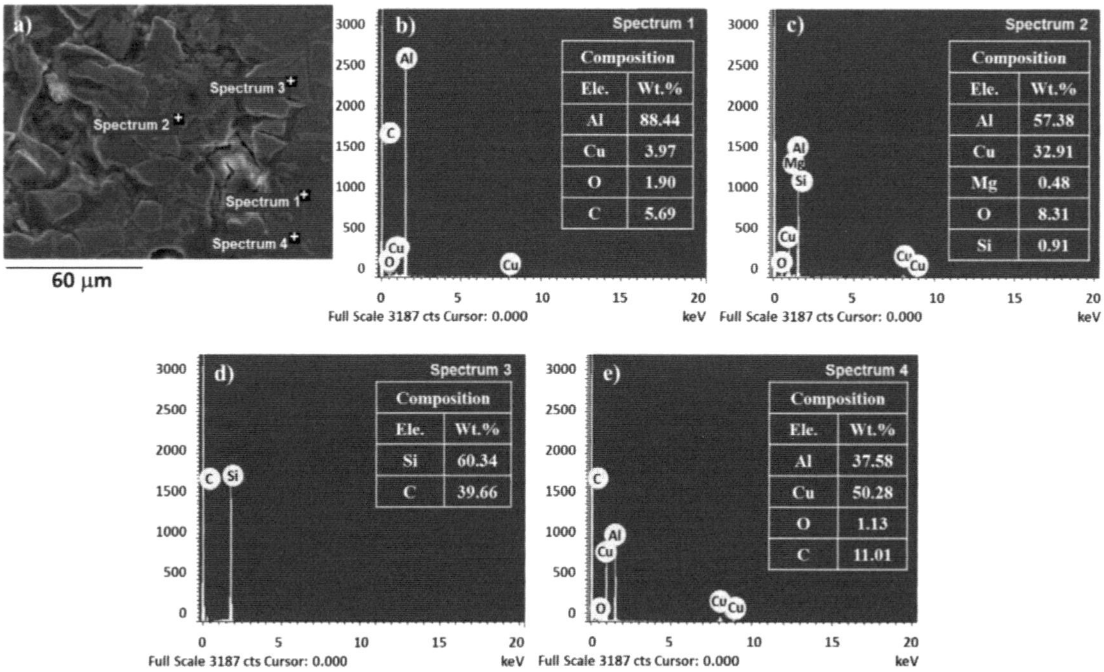

Figure 7. EDS point analysis of 40 vol.% SiC_p-reinforced aluminium composite after age hardening.

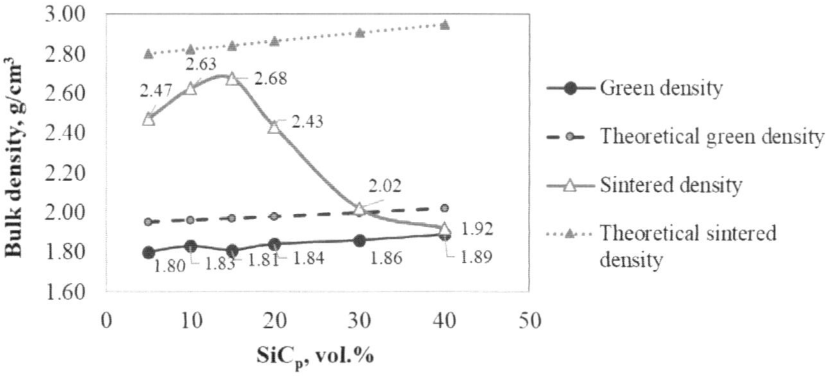

Figure 8. Measured and theoretical density of 5 - 40 vol.% SiC_p reinforced aluminium composite.

3.4. Green and sintered density

Green and sintered density of the feedstocks and composite after liquid phase sintering is illustrated in Fig. 8. A slight increasing trend of the feedstocks' green density is owning to SiC_p addition. According to the rule of mixture and by taking the feedstocks prepared at 55% solid loading, density of aluminium alloy (ρ_{Al} = 2.78 g.cm^{-3}), density of SiC_p (ρ_{SiC} = 3.20 g.cm^{-3}) and density of multi - component binder (ρ_{Binder} = 0.94 g.cm^{-3}), the theoretical green density of the feedstocks is found in a range of 1.95 - 2.02 g.cm^{-3} for 5 - 40 vol.% SiC_p additions respectively. The measured green density of the feedstock is 1.80-1.89 g.cm^{-3} for 5 - 40 vol.% SiC_p additions. Inter-particle porosity especially along SiC_p clusters might have been the reason, which accounts for 91.9 - 93.6% theoretical density.

In contrast to green density reported earlier, the sintered density of the composite shows the optimum values at 2.68 g.cm^{-3} for the composite reinforced with 15 vol.% SiC_p addition. This accounts for 94.2% theoretical sintered density at the optimum value. Beyond this range, an adversative role of inter - particle porosity along the SiC_p clusters has interfered, as shown previously. The minimum values of 65.1% theoretical sintered density is for composite reinforced with 40 vol.% SiC_p addition. A significant reduction in sintered density at high SiC_p addition is due to poor particle - packing among Al - Al powders, Al powder - SiC_p and SiC_p - SiC_p, which will be later discussed.

3.5. Macro vickers hardness

Aluminium composite of 5 - 40 vol.% SiC_p additions in the as - sintered and age - hardened conditions exhibits a similar trend on hardness result to that of sintered density, giving the maximum Vickers hardness at 138.9 H_v and 166.6 H_v for 15 vol.% SiC_p addition in the as - sintered and age - hardened conditions respectively, as shown in Fig. 9. It is believed that particle - packing among aluminium and SiC_p played a similar role to that of sintered density, but in a lesser extent due partly to greater hardness of SiC_p resisting the hardness indentation.

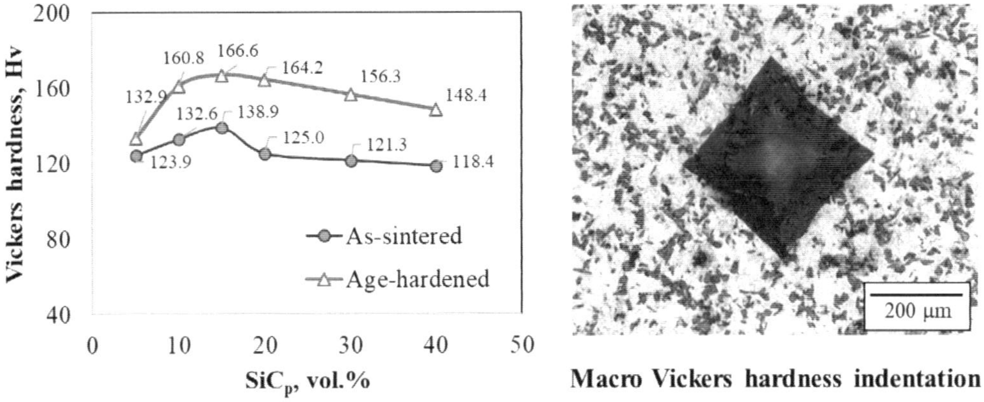

Figure 9. Macro Vickers hardness of 5 - 40 vol.% SiC_p reinforced aluminium composites.

4. Discussion

4.1. Phase evolution during liquid phase sintering

Aluminium composite has shown to experience liquid phase sintering, indicated by close Al - SiC_p and Al - Al interfaces especially for 5 - 20 vol.% SiC_p additions. The penetration of the aluminium along grain boundaries (less spherical grains) signifies good wettability between two adjacent interfaces [5]. The composite having 30 - 40 vol.% SiC_p addition however possessed very poor interfaces due to SiC_p clustering. Discussion on phase evolution during liquid phase sintering is now on i) major phase evolution during liquid phase sintering and ii) AlN formation.

First, phase evolution as evidenced from XRD, SEM and EDS analyses confirmed the Al - Cu alloy remained as the matrix with smaller SiC_p distributed throughout. A greater degree of SiC_p clustering was obtained at higher SiC_p addition. According to Al - Cu equilibrium phase diagram, the composite containing 4.77 wt.% Cu will experience the first formation of the liquid phase around 570 °C, and increasing liquid amount is expected towards 650 °C in the semi - solid regime. The DSC results also indicated the same range according to the endothermic reactions. Since the aluminium alloy - SiC_p composite is nonetheless considered as a non - equilibrium system and research by S. Wu, et. al., suggested the addition of SiC_p reinforcement increased both liquidus temperature (T_L) and eutectic temperature (T_E), depending on the amount of reinforcement [6]. In this research work, though the sintering temperature of 645 - 660 °C is exceeding that suggested from the Al - Cu equilibrium phase diagram, the solid - liquid range might be presumed according to the non - equilibrium system. As the higher temperature proceeds, higher content of the liquid phase promotes greater diffusion favorable for grain rearrangement; hence, better particle - packing, especially for the composite possessing low SiC_p content of 5 - 20 vol.%. It is explained that as the aluminium liquid melt spread on the SiC_p surface, the sintered bond was dissolved, and the inter - particle pores, if smaller than that of the matrix grains, would be annihilated due to the capillary force to pull the solid grains together [5]. In addition to those aluminium matrix and reinforcement, minor phases such as Al_2Cu and Mg_2Si are recognized, confirmed by XRD and SEM + EDS analyses, both in the as - sintered and age - hardened conditions. Much finer precipitates after age - hardening therefore reveal less visible XRD spectra as illustrated in Fig. 4 b).

Secondly, though AlN are not the major phase (neither matrix nor the reinforcement), AlN spectra are however obvious from the XRD result. Since AlN is considered as a significant phase in promoting densification [7] and mechanical properties, discussion is then given herein. Evidence of exothermic reactions taking place at 458 - 461 ^oC (DSC4), corresponding to the weight gains on the TG curves at 471 - 479 ^oC (TG4) might possibly be related to the AlN formation. Several research indicated the possibility of AlN formation at temperatures as low as 530 - 540 ^oC [8, 9], by direct nitridation or self - gettering. It was explained that the Al_2O_3 layer on the aluiminium powder was reduced by magnesium [10]. The Al_2O_3 layer was then consumed to disclose the aluminium, which is readily reacted with the N_2 within the sintering atmosphere to finally obtain AlN. The possible reactions proposed are as follows [7, 10]

$$3Mg + Al_2O_3 \rightarrow 3MgO + 2Al \quad (1)$$

or

$$3Mg + 4Al_2O_3 \rightarrow 3MgAl_2O_4 + 2Al \quad (2)$$

$$2Al + N_2 \rightarrow 2AlN \quad (3)$$

Alternatively, research by S.-H. Kim, et. al, proposed two steps of nitridation: i) the nitridation of aluminium via the native oxide layer in eqs. 4 - 5) and ii) direct nitridation of aluminium melt [11].

$$Al_2O_3\ (s) + N_2\ (g) \rightarrow 1AlON\ (s) + \tfrac{1}{2}\, O_2\ (g) \quad (4)$$

$$xAl_2O_3\ (s) + Al\ (l) \rightarrow yAlN\ (s) + O_2\ (g) \quad (5)$$

Their research suggested the nitridation around 620 ^oC where the weight gain and exothermic reaction were detected. T. Okada, et. al, [8] on the other hand, suggested the temperature as low as 520 ^oC is for partial nitridation and where complete direct nitridation was obtained at 540 ^oC, depending on the compact pressure employed. However, P. Sthapitanonda and J. L. Margrave [12] suggested the initial nitridation at 475 ^oC detected by weight gain at 0.10 $\mu g/cm^2$. In this research, the nitridation temperature is recognized by the weight gain around 472 ^oC. It is proposed here that with the addition of hard/abrasive SiC_p, such Al_2O_3 layer might have been disrupted during ball-mill mixing and PIM process [13]. This in turn offers a fast track for Mg to migrate through such oxide layer to reduce the Al_2O_3 film at low temperature as low as 300 - 500 ^oC [14].

Thirdly, if reactions towards the liquid phase is now considered, the weight gains detected on TG curves (TG5), showing the onsets at 643.7 and 847 ^oC for the feedstocks having 10 and 40 vol.% SiC_p addition for comparison. Though it is known that SiC is readily attacked by the molten aluminium to give Al_3C_4 [15] via dissolution and precipitation in 657 - 827 ^oC temperature range (eq. 6), no Al_3C_4 was detected by XRD results in this research. This is owing to peroxidation of SiC_p at 1000 ^oC/4 h prior to feedstock preparation to give the protective covering SiO_2 film (eq. 7) against the Al_3C_4 formation during liquid phase sintering [16, 17].

$$4Al\ (l) + 3SiC\ (s) \rightarrow Al_4C_3\ (s) + Si\ (s) \quad (6)$$

$$2SiC\ (s) + 3O_2\ (g) \rightarrow 2SiO_2\ (g) + 2CO\ (g) \quad (7)$$

Such SiO_2 film can be attacked by aluminium melt to form elemental silicon via redox reaction, which later diffuses to the aluminium melt whereas aluminium will diffuse in the counter direction towards SiO_2, giving the Al_2O_3 formation (eq. 8) as proposed by M. Yan and Z. Fan [18]. Evidence of free Si and Al_2O_3 was also detected via XRD result.

$$4Al\ (l) + 3SiO_2\ (s) \rightarrow 3Si\ (l) + 2Al_2O_3\ (s) \quad (8)$$

In summary, liquid phase sintering of SiC_p - aluminium composite involved multiple solid phases, which have different solubility-temperature-relations. Phase evolution during liquid phase sintering of PIMed SiC_p – reinforced aluminium composite is then proposed as depicted in Fig. 10 (adopted from Randall M. German [5]). First, i) solid state sintering via Al_2O_3 surface film reduction by magnesium before entering the liquid phase sintering. As preferential dissolution of solids by the liquid melt during heating is at the grain boundaries [19], it might be that after the aluminium melt protruded out from the disrupted Al_2O_3 surface film, AlN formation then progressed as aluminium melt content increased with increasing temperature. In the presence of magnesium, wettability was enhanced and promoted ii) liquid flow and rearrangement of the solid grains. Then iii) solution - reprecipitation took place plausibly by the aluminium melt, acting as the transient liquid, to dissolve smaller aluminium solid grains and reprecipitated in the tension area of larger grains. Aluminium melt attack on SiO_2 film might also be recognized along with continuous AlN formation during this stage before proceeding to the final densification by iv) solid phase sintering. Densification of the composite was thus enhanced along the course of liquid phase sintering.

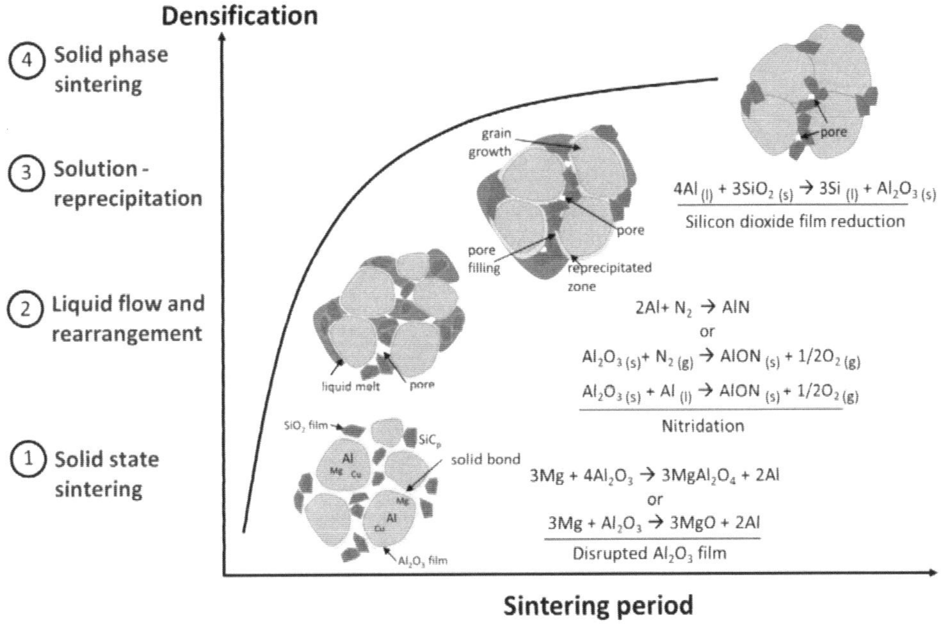

Figure 10. Schematic of microstructure evolution in liquid phase sintering of PIMed SiC_p - reinforced aluminium composite.

4.2. Effects of SiC_p addition on properties of PIMed aluminium composite

Increasing SiC_p content to 30 - 40 vol.% increased the solid fraction that required greater heat mass due to SiC_p's refractoriness. High surface energy and angular shape of SiC_p resulted in clustering which increased unreacted SiC_p - SiC_p interface. This poor interface cannot be wet by aluminium melt during liquid phase sintering, leading to excessively large pores. If pores are larger than the grain size, in the case of the composite having high SiC_p content of 30 - 40 vol.%, they are difficult to be filled by the liquid melt and required considerable time. Unfortunately, if there is any gas trapped in the pores, such as O_2 generating during AlN formation, then the combined effects of gas pockets in the pores made them spherical and difficult to collapse, which prohibited the final densification. This led to reduced density and hardness of the composite at higher SiC_p addition of 30 - 40 vol.%. If pores are smaller than the solid particles (Al powder and SiC_p), they will be later filled with the liquid aluminium due to the capillary force. Shrinkage then resulted and this is the case for the composite having 5 - 20 vol.% SiC_p addition, where volumetric shrinkage was reported to be 35 - 40% [20].

5. Conclusions

1) The PIMed 5 - 40 vol.% SiC_p reinforced aluminium composite experienced liquid phase sintering over the investigated sintering temperatures, suggested by thermal analysis.
2) Aluminium composite at 5 - 20 vol.% SiC_p addition exhibited close Al - Al and Al - SiC_p interfaces indicating liquid flow and rearrangement of the solid grains, leading to densification and shrinkage. However, high SiC_p addition at 30 - 40 vol.% resulted in large pores along SiC_p clustering, prohibited pore filing and wetting.
3) The optimum SiC_p addition is 15 vol.% to offer the maximum bulk density at 2.68 g.cm^{-3} at 94.2 % theoretical density and the maximum macro Vickers hardness at 138.9 and 166.6 H_v in the as - sintered and age - hardened respectively.
4) AlN was detected during liquid phase sintering, which improved densification and hardness.

References

[1] H. Abdoos, H. Khorsand, and A.A. Yousefi: Int. J. Mater. Res. Vol. 108 (2017) p. 237-244.
[2] H. Ye, X.Y. Liu, and H. Hong: J. Mater. Process Tech. Vol. 200 (2008) p. 12-24.
[3] V. G. Dhore, W.S. Rathod, and K.N. Patil: Mater. Today Proc. Vol. 5 (2018) p. 20690-20698.
[4] S. G. Lamarre, V. Demers, and J.-F. Chatelain: Int. J. Adv. Manuf. Tech. Vol. 91 (2017) p. 2595–2605.
[5] R.M. German, in: Sintering: From empirical observations to scientific principles, Publications/Elsevier (2014) in press.
[6] S. Wu, P. An, Y. You, T. Kanno, and H. Nakae: in *13th International Conference on Composite Materials (ICCM13)* (2001) Beijing, China.
[7] T. W. Kim, Mater. Sci. Eng., A: Vol. 483-484 (2008) p. 648-651.
[8] T. Okada, M. Toriyama, and S. Kanzaki: J. Mater. Sci. Vol. 35(2000) p. 3105-3111.
[9] G. B. Schaffer, and B.J. Hall, Metall. Mater. Trans. A. Vol. 33 (2002) p. 3279-3284.
[10] S. Zhongliang, G. Mingyuan, L. Junyou, L. Guoquan, L. Jae-chul, Z. Di, W. Renjie: Chin Sci Bull. Vol. 46 (2001) p. 1984-1952.
[11] S. H. Kim, J. H. Noh, J. P. Ahn, J.-C. Lee, H. Kwon, J. Lee, H. R. Yang, and K. B. Lee: Metall. Mater. Trans. A, Vol. 46 (2015) p. 496-504.
[12] P. Sthapitanonda and J. L. Margrave: J. Phys. Chem. Vol. 60 (1956) p. 1628-1633.
[13] M. P. Thomas, and J.E. King: J. Mater. Sci. Vol. 29 (1994) p. 5272-5278.
[14] A. Kimura, M. Shibata, M. Katayama, T. Kanie, and H. Takada: Appl. Phys. Lett. Vol. 70 (1997) p. 3615-3617.
[15] J. C. Viala, F. Bosselet, V. Laurent, and Y. Lepetitcorps: J. Mater. Sci., Vol. 28 (1993) p. 5301-5312.
[16] J. Roy, S. Chandra, S. Das, and S. Maitra: Rev. Adv. Mater. Sci. Vol. 38 (2014) p. 29-39.
[17] A. Ureña, E. E. Martínez, P. Rodrigo, and L. Gil: Compos. Sci. Technol. Vol. 64 (2004) p. 1843-1854.
[18] M. Yan, and Z. Fan: J. Mater. Sci. Vol. 63 (2001) p. 285-295.
[19] R. M. German, P. Suri, and S.J. Park: J. Mater. Sci. Vol. 44 (2009) p. 1-39.
[20] T. Patcharawit, A. Ngeekoh, U. Kitkhamthorn, and N. Chuankrekkul: in *2019 4th International Conference on Design (D2ME 2019)* Phuket, Thailand.

IOP Conf. Series: Materials Science and Engineering **600** (2019) 012015 doi:10.1088/1757-899X/600/1/012015

Effect of Na_2SiO_3/NaOH ratio on compressive strength of lateritic soil mixed with bagasse ash geopolymer

Danai Jongphimai[1], Cherdsak Suksiripattanapong[2] and Jaksada Thumrongvut[3]

Department of Civil Engineering, Faculty of Engineering and Architecture, Rajamangala University of Technology Isan, Nakhon Ratchasima, Thailand

corresponding author's e-mail: cherdsak.su@rmuti.ac.th

Abstract. This paper investigated the effect of Na_2SiO_3/NaOH ratio on compressive strength of lateritic soil-bagasse ash geopolymer. Lateritic soil was obtained from Sikhio District, Nakhon Ratchasima. Bagasse ash is a waste from the burning of sugarcane for electricity The lateritic soil to binder ratio was fixed at 85/15. The alkaline activator was composed of sodium silicate solution (Na_2SiO_3) and sodium hydroxide solution (NaOH) at concentration of 5 M. The Na_2SiO_3/NaOH ratios were 80/20, 70/30 and 50/50. The compressive strength of lateritic soil-bagasse ash geopolymer samples was evaluated at curing time of 7, 14, 28 and 60 days. It can be seen that the compressive strength of lateritic soil mixed with bagasse ash geopolymer increases as the Na_2SiO_3/NaOH ratio and curing time increase. The maximum compressive strength of lateritic soil-bagasse ash geopolymer was found at the Na_2SiO_3/NaOH ratio of 80/20 and curing time of 60 days, which gave maximum compressive strength of 9.84 MPa.

1. Introduction

Interlocking brick has been widely used construction material. The raw material used in the production of interlocking blocks consists of soil, cement and water. Generally, interlocking brick was produced by compression machine. However, interlocking brick used Portland cement. The process of cement releases large amount of carbon dioxide into the atmosphere and causes the greenhouse gases [1,2]
Geopolymerization is a process that creates cementitious compounds by the dissolution of aluminosilicate compounds by highly alkaline liquids. The geopolymerization of aluminosilicate-rich materials results in strong cementitious compounds [3-7]. Bagasse ash is a waste from the burning of sugarcane for electricity. Particles of bagasse ash are large and high porosity. Bagasse ash has main aluminosilicate-rich material which can be classified as Class N pozzolanic materials according to ASTM C 618 [8].

This research aimed to investigate the compressive strength of lateritic soil mixed with cement and bagasse ash. Alkaline activator, L consists of sodium silicate solution (Sodium silicate solution, Na_2SiO_3) and sodium hydroxide solution (Sodium hydroxide solution, NaOH). The strength of sample was evaluated using the unconfined compressive strength (UCS) at curing time of 7, 14, 28 and 60 days.

2. Materials and testing methods

2.1. Materials

Lateritic soil (LS) was obtained from Sikhio District, Nakhon Ratchasima. The specific gravity of LS is 2.72. the liquid limit and the plastic limit of LS are 32 and 21, respectively. This soil can be classified as clayey sand (SC) according to the Unified Soil Classification System (USCS).
Bagasse ash (BA) is a waste from the burning of sugarcane for electricity, the Khonburi sugar factory. The chemical composition of bagasse ash was shown in Table 1. It can be seen that the BA has SiO_2 of 70.91 and Al_2O_3 of 4.28, respectively. BA can classified as type N according to ASTM C 618.
The alkaline solution consists of sodium hydroxide solution (NaOH) with a concentration of 5 molar and sodium silicate solution (Na_2SiO_3).

(a) Lateritic soil

(b) Bagasse ash

Figure 1. Characteristic of lateritic soil and bagasse ash

2.2. Sample preparation

The lateritic soil-bagasse ash geopolymer is a mixture of laterite, bagasse ash, and liquid alkaline activator (L) (sodium hydroxide solution, NaOH and sodium silicate solution, Na_2SiO_3). Lateritic soil and bagasse ash ratio was fixed at 85/15. The Na_2SiO_3/NaOH ratios were 80 / 20, 70/30 and 50/50. NaOH concentration was fixed at 5 molar. Modified Proctor compaction effort in accordance with

ASTM D1557-12 [9] was used to compact the lateritic soil-bagasse ash geopolymer. Next the lateritic soil-bagasse ash geopolymer was prepared by static compression with a 50 mm in diameter and a 100 mm in height using a cylindrical split mold. Then, the specimens were wrapped in clear vinyl to prevent moisture loss throughout curing. These specimens were subjected to Unconfined compressive strength (UCS) tests with reference to ASTM D2166 [10] at curing time of 7, 14, 28, and 60 days.

Table1. Chemical compositions of BA.

Chemical compositions	BA (%)
SiO_2	70.91
Al_2O_3	4.28
Fe_2O_3	6.61
CaO	8.91
MgO	0.75
SO_3	0.56
Na_2O	0.11
K_2O	3.48
MoO_3	N.D.
LOI	3.28

3. Research results and discussion results

Figure 1 shows the relationship between relationship between dry weight unit and liquid alkaline content at different $Na_2SiO_3/NaOH$ ratios. It was found that The $Na_2SiO_3/NaOH$ ratios have an effect on the dry weight unit of lateritic soil-bagasse ash geopolymer. The dry weight unit of lateritic soil-bagasse ash geopolymer increased with increasing $Na_2SiO_3/NaOH$ ratios. The optimum liquid alkaline content of lateritic soil-bagasse ash geopolymer was 13, 14 and 15 % for $Na_2SiO_3/NaOH$ ratios of 80/20, 70/30 and 50/50, respectively. The maximum dry weight unit of lateritic soil-bagasse ash geopolymer was 18.5 kN/m^3 at $Na_2SiO_3/NaOH$ ratios.

Figure 2 shows the compressive strength of lateritic soil-bagasse ash geopolymer at different $Na_2SiO_3/NaOH$ ratios. The compressive strength of lateritic soil-bagasse ash geopolymer increases with increasing $Na_2SiO_3/NaOH$ ratios. The maximum compressive strength of the lateritic soil-bagasse ash geopolymer sample was found at $Na_2SiO_3/NaOH$ ratio of 80/20 and curing time of 60 days, which gave maximum compressive strength of 9.84 MPa. The compressive strength of lateritic soil-bagasse ash geopolymer increases as curing time increases until the curing time of 28. Then the compressive strength of lateritic soil-bagasse ash geopolymer is constant as curing time increases. It is because the alkaline solution can leach more silica and alumina from bagasse ash and result in the formation of geopolymerization products.

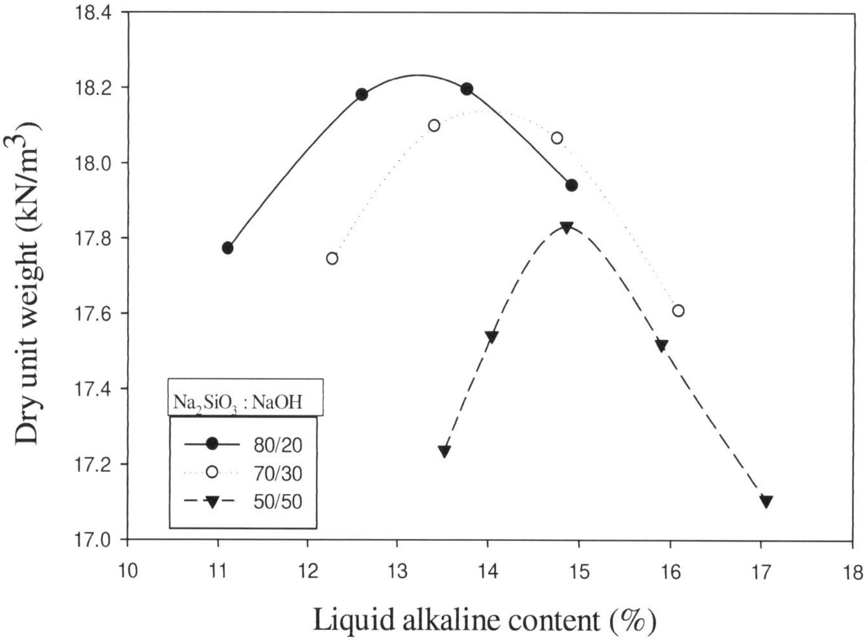

Figure 2. Relationship between dry weight unit and liquid alkaline content at different Na_2SiO_3/NaOH ratios.

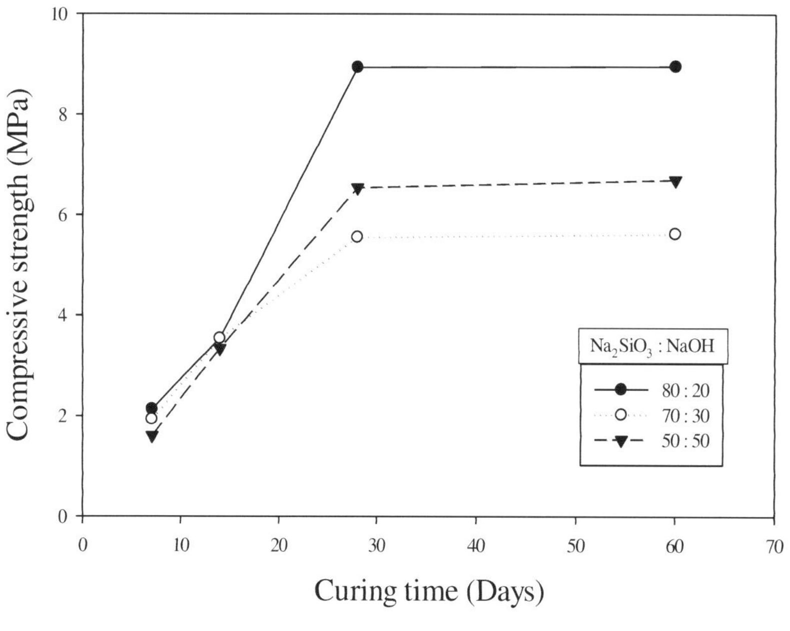

Figure 3. The compressive strength of compressive strength of lateritic soil-bagasse ash geopolymer at different Na_2SiO_3/NaOH ratios

4. Conclusions

The effect of Na_2SiO_3/NaOH ratio on compressive strength of lateritic soil-bagasse ash geopolymer was evaluated in this study. The conclusions can be drawn as follows. The dry weight unit of lateritic soil-bagasse ash geopolymer increased with increasing Na_2SiO_3/NaOH ratios. The optimum liquid alkaline content of lateritic soil-bagasse ash geopolymer was 13, 14 and 15 % for Na_2SiO_3/NaOH ratios of 80/20, 70/30 and 50/50, respectively. The maximum compressive strength of the lateritic soil-bagasse ash geopolymer sample was found at Na_2SiO_3/NaOH ratio of 80/20 and curing time of 60 days, which gave maximum compressive strength of 9.84 MPa.

Acknowledgments

This work was financially supported by the Thailand Research Fund under the Research and Researcher for Industry (RRi) program Grant No. RTA MSD61I0004. The authors acknowledge the financial support from the Pakpanyawat Ltd.,Part and the Rajamangala University of Technology Isan.

References

[1] Suksiripattanapong, C., Horpibulsuk, S., Chanprasert, P., Sukmak, P. and Arulrajah, A. (2015), “Compressive strength development in fly ash geopolymer masonry units manufactured from water treatment sludge”, Construction and Building Materials.

[2] Suksiripattanapong C., Kua T., Arulrajah A., Maghool F.and Horpibulsuk S., Strength and microstructure properties of spent coffee grounds stabilized with rice husk ash and slag geopolymers, Construction and Building Materials

[3] Davidovits, J., (1999). Chemistry of geopolymer system, Terminology. In Proceeding of Second International Conference Geopolymer’99, France. : pp 9–39.

[4] Suman Saha, C. Rajasekaran., (2017). Enhancement of the properties of fly ash based geopolymer paste by incorporating ground granulated blast furnace slag Enhancement of the properties, Construction and Building Materials.

[5] Aleksandar Nikolov, Ivan Rostovsky, Henk Nugteren., (2017). Geopolymer materials based on natural zeolite, Case Studies in Construction Materials

[6] Jamsawang P., Poorahong H., Yoobanpot N., Songpiriyakij S. And Jongpradist P., (2017), Improvement of soft clay with cement and bagasse ash waste., Construction and Building Materials

[7] Aleksandar Nikolov, Ivan Rostovsky and Henk Nugteren, (2017), Geopolymer materials based on natural zeolite, Case Studies in Construction Materials

[8] ASTM C618, 2008. Standard specification for coal fly ash and raw or calcined natural pozzolan for use in cement. Annual Book of ASTM Standard. Vol.04.01

[9] ASTM D1557. Standard Test Methods for Laboratory Compaction Characteristics of Soil Using Modified Effort (56,000 ft-lbf/ft^3 (2,700 kN-m/m^3))

[10] ASTM D2166. Standard Test Method for Unconfined Compressive Strength of Cohesive Soil

Analysis of fire-damaged concrete-filled steel tube columns confined with FRP sheets

Dat Thanh Vu and Akhrawat Lenwari

Composite Structures Research Unit, Department of Civil Engineering, Faculty of Engineering, Chulalongkorn University, Phayathai Road, Pathumwan, Bangkok 10330, Thailand.

Akhrawat.L@chula.ac.th

Abstract. This paper presents an analysis of fire-damaged circular concrete-filled steel tube (CFST) stub columns confined with fiber-reinforced polymer (FRP) sheets under compression. Three main components in the analysis include (1) 2D heat transfer analysis to predict the temperature distributions in the CFST cross section under ISO-834 standard fire, (2) Post-fire mechanical properties of steel and concrete, and (3) Analytical model that incorporates the confinement effect to predict the load capacity of FRP-confined fire-damaged CFST columns. After validated with existing experimental results, the proposed analysis is employed to investigate the effects of heating period and number of FRP layers on the effectiveness of the FRP confinement method.

1. Introduction

Concrete-filled steel tube (CFST) columns offer several advantages due to composite action including reduced self-weight, fast construction, and high fire resistance [1]. With increasing adoption of CFST in engineering practice, the fire resistance and post-fire reparability of CFST columns are important for fire safety purpose [2]. Among several strengthening methods for CFST columns, the use of external wraps such as fiber-reinforced polymer (FRP) sheets is promising due to high strength-to-weight ratio, high corrosion resistance, and ease of installation [2]. So far, however, there have been few limited analytical studies on the fire-damaged CFST columns confined with FRP sheets [2, 3].

In this paper, the analysis of axial compression behavior of fire-damaged CFST columns confined with FRP sheets is presented. First, the 2D finite element heat transfer analysis is carried out to obtain the temperature profile (time-temperature relationship) of each discretized CFST cross-section layer in which temperature is assumed to be uniform. Next, the post-fire mechanical properties of steel and concrete proposed by Han and Huo [4] and Chang et al. [5] are adopted. Finally, by adopting the model of confined CFST columns [1] with post-fire material properties, the axial load capacities of FRP-confined fire-damaged CFST columns are calculated by the layer discretization method. The experimental results by Tao, Han [2] are used to verified the analysis. Based on the proposed analysis, a parametric study is performed to investigate the effects of number of FRP layers and heating period on the strength of FRP-confined fire-damaged CFST columns.

IOP Conf. Series: Materials Science and Engineering **600** (2019) 012016 doi:10.1088/1757-899X/600/1/012016

2. Analysis of FRP-confined fire-damaged CFST columns

2.1. Heat transfer analysis

2.1.1. Finite element model. A commercial finite element analysis package ANSYS is used to perform the heat transfer analysis to predict the temperature distributions within the CFST cross section during fire exposure (heating and cooling phases). The applied thermal boundary conditions and thermal parameters of concrete and steel are defined according to the recommendations from Eurocode 4 [6]. For the heating phase, the loading temperature refers to ISO 834 standard time-temperature curve:

$$T = T_0 + 345 \log_{10}(8t + 1) \tag{1}$$

where T_0 is the ambient temperature and T is the elevated temperature at time t (min).

2.1.2. Experimental results for model verification. To examine the validity of the 2D heat transfer analysis, an experimental program is conducted. Circular CFST tubular columns were heated for 120 min under ISO 834 standard fire inside the fire testing furnace, as shown in Figure 1a. Figure 1b shows a good agreement between predicted temperatures and experimental data. The maximum differences are 2.5% and 4.5% for peak steel tube and concrete temperatures, respectively.

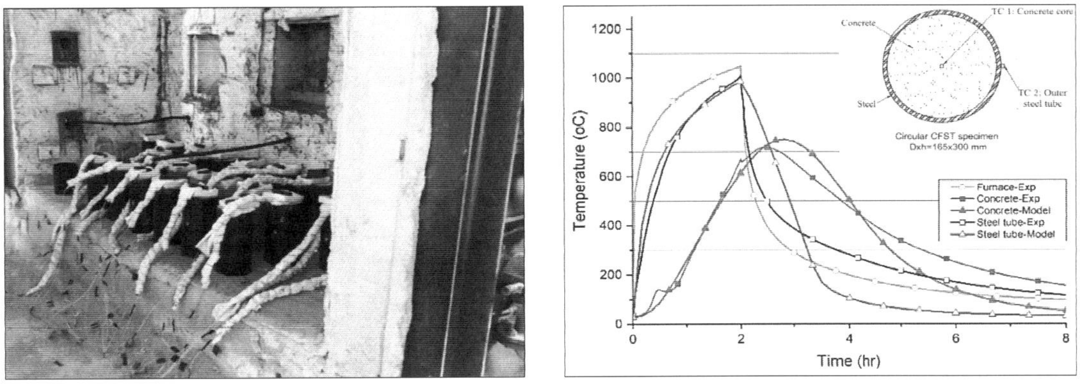

(a) (b)

Figure 1. Experimental program: (a) Fire testing furnace and CFST specimens; (b) Comparison between experimental results and heat transfer analysis predictions.

2.2. Post-fire material properties

The post-fire mechanical properties of steel and concrete proposed by Han and Huo [4] and Chang et al. [5] are adopted for prediction of compression behavior of CFST columns. The post-fire properties of steel and concrete are assumed to depend on the peak temperature obtained from the heat transfer analysis. The accuracy of the material model has been validated in Rush et al. [7].

2.3. Confinement effect by FRP

By adopting the post-fire material properties for the external confinement effect model, the compression behavior of confined fire-damaged CFST columns can be determined. In this paper, steel and concrete refer to the post-fire condition unless specified. Lai and Ho [1] proposed the relationship between axial strain (ε_z) and hoop strain (ε_θ) of the CFST column as follows:

$$\varepsilon_z = LS\left(\frac{f'_c}{30}\right)^m \left\{\varepsilon_{co}\left[1 + 0.75\left(\frac{-\varepsilon_\theta}{\varepsilon_{co}}\right)\right]^{0.7} - \varepsilon_{co}\exp\left[7\left(\frac{\varepsilon_\theta}{\varepsilon_{co}}\right)\right] + 0.07(-\varepsilon_\theta)^{0.7}\left[1 + 26.8\left(\frac{f_r}{f'_c}\right)\right]\right\} \tag{2}$$

where LS is the parameter reflecting the effect of external confinement. For the FRP-confined column, LS=0.665; m is the parameter considering effect of concrete strength; ε_{co} is the strain corresponding to unconfined concrete strength; f_c' is the compressive strength of unconfined concrete; f_r is the confining pressure from both steel tube (f_{rS}) and FRP confinement:

$$f_r = f_{rS} + f_{rE} \tag{3}$$

$$f_{rS} = -\frac{2t}{D_0 - 2t}\sigma_{s\theta} \tag{4}$$

$$f_{rE} = -\frac{2t_{FRP}}{D_0 - 2t}\sigma_E \tag{5}$$

where $\sigma_{s\theta}$ is the hoop stress provided by the steel tube, which can be evaluated by Hooke's Law, the yield surface is determined by Von Mises yield criterion, and Prantld-Reuss's theory [1]; t and t_{FRP} are thicknesses of steel tube and FRP wrap, respectively; σ_E is the stress provided by external confinement (FRP):

$$\sigma_E = \begin{cases} \varepsilon_{ssE}E_{ssE} & E_{ssE} \le \sigma_{ssE} \\ \varepsilon_{ssE}\sigma_{ssE} & E_{ssE} > \sigma_{ssE} \end{cases} \tag{6}$$

where ε_{ssE}, E_{ssE}, and σ_{ssE} are the average hoop strain, elastic modulus, and yield stress of FRP, respectively. Assume that $\varepsilon_{ssE} = \varepsilon_\theta$.

For the confined core concrete, the stress-strain equation is:

$$\frac{f_{cc}}{f_{ccp}} = \frac{A\left(\frac{\varepsilon_z}{\varepsilon_{cc}}\right) + B\left(\frac{\varepsilon_z}{\varepsilon_{cc}}\right)^2}{1 + (A-2)\left(\frac{\varepsilon_z}{\varepsilon_{cc}}\right) + (B-1)\left(\frac{\varepsilon_z}{\varepsilon_{cc}}\right)^2} \tag{7}$$

$$\varepsilon_{cc} = \varepsilon_{co}\left[1 + \frac{(17 - 0.06f_c')f_r}{f_c'}\right] \tag{8}$$

$$\frac{f_{ccp}}{f_c'} = 1 + 4.1\left(\frac{f_r}{f_c'}\right) \tag{9}$$

where A and B are parameters that govern the shape of stress-strain curve [8]; f_{cc} is the confined concrete stress; f_{ccp} and ε_{cc} are the peak confined concrete stress and corresponding axial strain.

2.4. Compressive strength of FRP-confined fire-damaged CFST columns

The composite cross section is divided into m layers. Each layer has its own peak exposed temperature and mechanical behavior. The steel tube is considered as one layer. The axial load capacity (F_t) is calculated as follows:

$$F_t = F_c + F_s \tag{10}$$

$$F_t = \sum_{i=1}^{m} F_{ci} + F_s \tag{11}$$

$$F_t = \sum_{i=1}^{m} f_{cci}A_{ci} + \sigma_{sz}A_s \tag{12}$$

where A_s and A_{ci} are areas of steel tube and confined concrete of layer i, respectively; σ_{sz} is the axial stress of steel tube which can be evaluated by Hooke's Law, and the aid of Prantld-Reuss's theory [1]; and f_{cci} is the stress of confined concrete of layer i.

2.5. Generation of the axial load-strain curves

Fig 2 shows an iterative process to generate the axial load-strain curves of the FRP-confined fire-damaged CFST column.

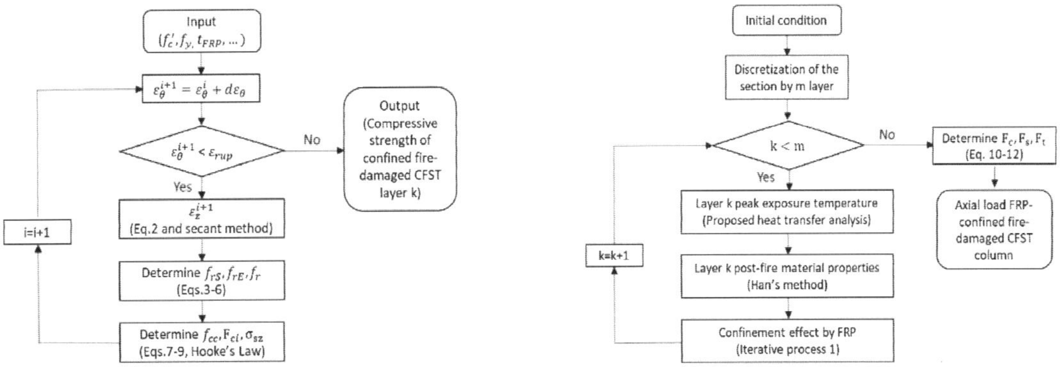

Figure 2. Iterative process: (a) Confinement effect by FRP (iterative process 1); (b) Generation of axial load-strain curve of FRP-confined fire-damaged CFST column (adapted from [1])

3. Verification

Table 1 shows a comparison between the experimental results by Tao et al. [2] and analysis results. Ratios of predicted-to-experimental load capacity range from 0.967 to 0.998. The close agreement implies that the proposed analysis can be used for predicting the compressive strength of FRP-confined fire-damaged CFST columns.

Table 1. Experimental results by Tao et al. [3] vs. analysis predictions

Specimen	D × t_s (mm × mm)	f_y (N/mm^2)	f_{cu} (N/mm^2)	Fire damage	Number of layers	Predicted load capacity (kN)	Experimental results (kN)	N_{pre}/ N_{Exp}
CSC	150 x 3.0	356	75	No	0	1911	1915	0.998
CSCF-0	150 x 3.0	356	75	ISO-3 h	0	910	915	0.994
CSCF-1	150 x 3.0	356	75	ISO-3 h	1	1003	1020	0.983
CSCF-2	150 x 3.0	356	75	ISO-3 h	2	1508	1560	0.967

Table 2. Material properties of CFRP and epoxy

Material	Elastic Modulus (kN/mm^2)	Tensile Strength (N/mm^2)	Ultimate Strain (%)	Thickness (mm)
CFRP	247	3950	1.60	0.170
Epoxy	2.7	42.8	1.85	-

4. Effects of parameters on compressive strength of FRP-confined fire-damaged CFST columns

Based on the proposed analysis, a parametric study is performed to investigate the effects of number of FRP layers and heating time on the compressive strength of CFST columns. In the study, specimen properties are taken from experiments by Tao et al. (Table 1 and 2). Fig 3 shows that the compressive strength of unconfined CFST columns decreases by 21%, 35% and 52% for 1-h, 2-h and 3-h ISO 834 standard fire exposures, respectively. Also, the two-layer FRP is more effective than one- layer FRP. For the two-layer FRP, the strength increases by 33%, 46%, 52%, and 66% for 0-h (no fire exposure),

1-h, 2-h, and 3-h ISO 834 standard fire exposures, respectively. For the one-layer FRP, the strength increases by 12%, 12%, 11%, and 10% for 0-hour (no fire exposure), 1-h, 2-h, and 3-h ISO-834 standard fire exposures, respectively. The confinement effectiveness of the two-layer FRP is higher for fire damaged CFST columns than undamaged ones, while the use of one-layer FRP provides similar confinement effectiveness.

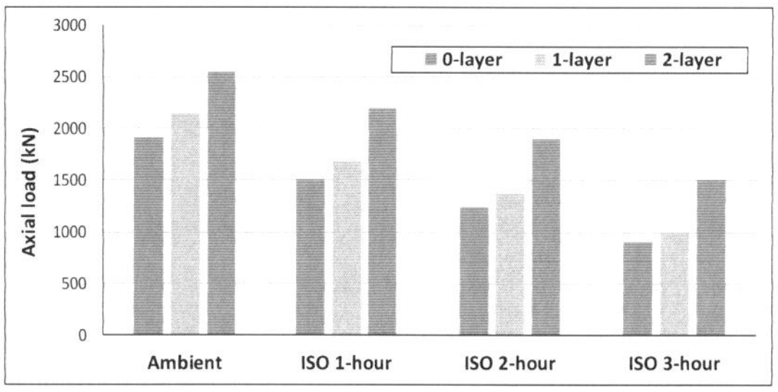

Figure 3. Effects of number of FRP layers and heating time parameters

5. Conclusions

In this paper, an analysis procedure for predicting the axial compression behavior of FRP-confined fire-damaged circular CFST columns is described. The 2D finite element model is used to perform the heat transfer analysis. The proposed analysis can accurately predict the temperature distribution within the CFST cross section during fire exposure and load carrying capacity of both unconfined and FRP-confined fire-damaged CFST columns. The parametric study shows that the confinement effectiveness depends on heating time (level of fire damage) and number of FRP layers.

Acknowledgments

Authors would like to acknowledge the financial support from the ASEAN University Network/Southeast Asia Engineering Education Development Network (AUN/SEED-Net) and technical assistance at the Fire Safety Research Centre (FSRC), Chulalongkorn University.

References

[1] Lai, M. and J. Ho, *A theoretical axial stress-strain model for circular concrete-filled-steel-tube columns.* Engineering Structures. Vol. 125 (2016). p. 124-143.
[2] Tao, Z., L.-H. Han, and L.-L. Wang, *Compressive and flexural behaviour of CFRP-repaired concrete-filled steel tubes after exposure to fire.* Journal of Constructional Steel Research. Vol. 63(8) (2007). p. 1116-1126.
[3] Wang, K., Y. Chen, J. Wan, S. Han, and L. Liao, *Compressive behavior of post-heated circular CFST short columns externally strengthened with CFRP sheets.* Journal of Adhesion Science and Technology. Vol. 32(8) (2018). p. 833-853.
[4] Han, L.-H. and J.-s. Huo, *Concrete-filled hollow structural steel columns after exposure to ISO-834 fire standard.* Journal of Structural Engineering. Vol. 129(1) (2003). p. 68-78.
[5] Chang, Y.-F., Y.-H. Chen, and M.-S. Sheu, *Residual stress–strain relationship for concrete after exposure to high temperatures.* Vol. 36(10) (2006). p. 1999-2005.
[6] Eurocode *4: Design of composite steel and concrete structures*. 2005, Part 1-2: General rules-structural fire design.
[7] Rush, D.I., L.A. Bisby, and A. Jowsey, *Residual capacity of fire-exposed concrete-filled steel hollow section columns.* Engineering Structures. Vol. 100 (2015). p. 550-563.
[8] Kwan, A., C. Dong, and J. Ho, *Axial and lateral stress–strain model for FRP confined concrete.* Engineering Structures. Vol. 99 (2015). p. 285-295.

IOP Conf. Series: Materials Science and Engineering **600** (2019) 012017 doi:10.1088/1757-899X/600/1/012017

Surface modification of activated carbon by surfactants mixtures

Olga Kochkodan[1,*], Victor Maksin[1], Nadiya Antraptseva[1], Viktor Kochkodan[2]

[1]National University of Life and Environmental Sciences of Ukraine, Kyiv, Ukraine
[2]Qatar Environment and Energy Research Institute, Hamad bin Khalifa University, Qatar Foundation, Doha, Qatar

[*]Corresponding author's email: okochkodan@hotmail.com

Abstract. Adsorption of surfactants is widely used for surface modification of different materials. Using of the mixtures of the surfactants in many cases is more attractive compared with single surfactants, however composition of mixed adsorption layers at solid surface was not sufficiently studied yet. In this work, the mutual effect of anionic and nonionic surfactants on the composition of the mixed adsorption layer on the surface of activated carbon (AC) was studied. Sodium hexadecyl sulphate (SHS) and oxyethylated octylphenol Triton X100 (TX-100) were used as anionic and nonionic surfactants, respectively. Mixed SHS/TX-100 systems have been studied over a concentration range of 0.1-8.0 mmol/L and the molar fractions of SHS in the mixtures were within 0.2-0.8. It was shown that the adsorption isotherms of SHS or TX-100 surfactants at AC could be describe by the Langmuir equation. For surfactants mixtures at low equilibrium concentrations, it was found that adsorption of SHS increases when TX-100 is present in the solution. This might be explained by the decreasing of electrostatic interactions between the head groups of SHS molecules in the mixed adsorption layers due to incorporation of the molecules of the nonionic surfactant. At higher total surfactant concentrations, SDS adsorbed at AC surface is replaced by TX-100 molecules and as result the nonionic surfactants is preferentially adsorbed on AC.

Keywords: Activated carbon; surface layer; sodium hexadecyl sulphate; Triton X100

1. Introduction

Adsorption of surfactants is widely used for surface modification of different materials in numerous industrial processes such as flotation, enhanced oil recovery, lubrication [1]. In many cases the surfactants mixtures are more beneficial than the single surfactants [1,2]. Because of broad usage, about 10 million ton of surfactants are produced annually [3]. As a result of the massive production the surfactants could enter and pollute surface and ground waters negatively affecting the environment [4].To optimize the practical application of the surfactants and their removal from water it is important to investigate the processes, which occur at the solid-liquid interface in the surfactants mixtures.

Activated carbon (AC) is widely used in different application and it is the most important sorbent for adsorptive removal of toxic pollutants from water [5]. Surface modification of the AC with surfactants, which is widely used to enhance removal of the contaminants, is based on deposition of surfactants nanolayers at AC [6]. The properties of the prepared composite AC/surfactants materials largely

depends on the structure of the adsorbed surfactants layers [1,6]. It should be mentioned that numerous studies on adsorption of different types of the surfactants from single surfactant aqueous solutions on AC and features of the adsorbed layers of single surfactants are available in the literature [1, 7-10], however only few papers are related to mixed surfactants systems on AC [11, 12]. Xiao et al. [11] studied adsorption of octyltriethylammonium bromide with sodium dodecylbenzenesulfonate and dodecylpyridinium chloride with sodium octanesulfonate mixed systems at AC. The authors reported that hydrophobic interactions between the surfactants and AC are mainly responsible for the adsorption of these compounds at the sorbent. It was also shown synergetic adsorption of the surfactants due to the electrostatic interaction between oppositely charged ions of the anionic and cationic surfactants [11]. On the other hand, Gao at al. [12] found that adsorption of the nonionic surfactant on AC decrease in the presence of cationic cetyltrimethylammonium bromide.
In this work, the adsorption of anionic sodium hexadecyl sulfate (SHS) and nonionic octylphenol ethoxylate Triton X100 (TX-100) surfactants from their mixed solutions has been studied. This work aims to provide better understanding of surfactant interactions at the surface of AC to optimize such interactions.

2. Materials and methods

AC of AG-3 type was used as a sorbent. To remove inorganic impurities AC was treated with 2M solutions of HNO_3 and with concentrated HF. Then the samples were washed with distilled water and dried at 105 °C. The total pore volume, volume of micropores, volume of mezopores and total surface area of AC were 2.9×10^{-4} m^3/kg, 2.6×10^{-4} m^3/kg, 3.0×10^{-5} m^3/kg and 6.7×10^{5} m^2/kg, respectively, as was evaluated by benzol adsorption. AC with a grain size of 0.50-0.63 mm was used in the experiments. SHS and TX-100 surfactants were obtained from Sigma-Aldrich (USA).
For adsorption experiments, weighted amounts (0.1-0.3 g) of AC were added to surfactant solutions. The prepared AC dispersions were left shacking for 40 h at 25 °C and then centrifuged at 4000 rpm for 10 min. The surfactant solutions were prepared at pH 6.0-6.5.
The adsorbed amount of the surfactant (A, mmol/g) was evaluated from the change in the surfactant concentrations as follows:

$$A = \frac{(C_0 - C_e)V}{m} \quad (1)$$

where C_0 and C_e are the surfactant concentrations (mmol/L) in the initial solution and in the equilibrium solution, respectively, V is the volume of the solution (L) and m is the AC loading in the solution (g).
TX-100 concentration in the solutions was analyzed spectrophotometrically by measuring the optical density of the solutions at 274 nm, while SDS content in the probes was evaluated by two-phase titration method [13].
The mole fractions of SHS surfactant on the AC surface and in the aqueous solutions were evaluated using the following equations:

$$\chi = \frac{A_{SHS}}{A_{SHS} + A_{TX-100}} \quad (2)$$

$$\alpha = \frac{C_{SHS}}{C_{SHS} + C_{TX-100}} \quad (3)$$

where, χ and α are the mole fractions of SHS on AC surface and in the mixed equilibrium solution, respectively, A_{TX-100} and A_{SHS} are the adsorption values for TX-100 and SHS surfactants at AC surface, C_{TX-100} and C_{SHS} are the concentrations of TX-100 and SHS surfactants in the mixed equilibrium solution.

The surface charge of the AC particles was evaluated with a Zeta-Meter System 3.0+ by using the Smoluchowski equation:

$$\zeta = \frac{\mu\eta}{\varepsilon} \quad (4)$$

where η is the zeta potential (mV), η is dynamic viscosity (Pa s) and ε is dielectric constant. The measurements were performed at the solution pH values of 6.0-6.5

3. Results and discussions

3.1. Adsorption of SHS and TX-100 from their single aqueous solutions

As seen in Fig. 1, the adsorption isotherm of SHS and TX-100 on AC surface follows the Langmuir type with a well-defined horizontal plateau in the neighborhood of their respective critical micelle formation concentration values.

The amounts of the adsorbed surfactants increase sharply with the surfactant concentration attaining at the equilibrium concentration of approximately 0.6 mmol/L for SHS (Fig. 1a) and 0.35 mmol/L for TX-100 (Fig. 1b). The strong adsorption of SHS and TX-100 is due to the hydrophobic interactions between the alkyl residue of the surfactants molecules and graphite lattice at AC surface, which are recognized as the main force involved in the surfactant adsorption at AC [7-9]. The maximum adsorbed amounts of SHS and TX-100 on AC are approximately 0.28 and 0.45 mmol/g, respectively.

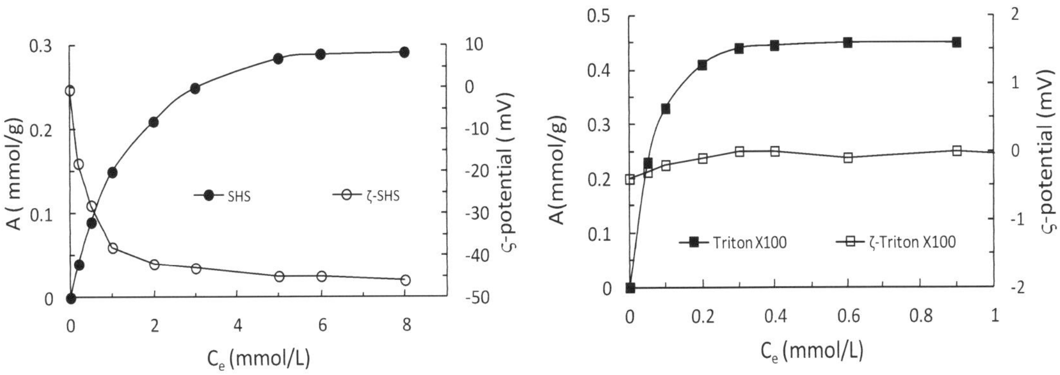

a) b)

Figure 1. Adsorption isotherms and zeta potentials of AC particles in single SHS (a) and TX-100 (b) surfactants solutions. pH 6.5.

As seen in Fig. 1, zeta potential of AC particles is very low (-0.4 mV) in an aqueous solution without the anionic surfactant. Similarly, Julien et al. [14] showed that the zeta potential of AC treated at 700 °C was close to zero. Adsorption of SHS molecules essentially changes zeta potential of AC particles even at low surfactant concentrations. Zeta potential values were observed to decrease steadily with the SHS concentration in the solutions, reaching a plateau value of about -45 mV (Fig. 1a). The change in the values of zeta potential of AC particles in SHS presence clearly indicates that significant adsorption of the surfactant at the AC surface take place. There is also a good correlation between the adsorbed amount of the anionic surfactant and values of zeta potential of the AC particles (Fig 1a). It might be assumed that SHS molecules adsorb with their hydrophobic residues oriented towards the AC surfaces while the anionic groups of the surfactant protrude into the aqueous phase. In contrast, as TX-100 is a nonionic surfactant its adsorption at the AC surface has practically no effect on ς-potential of AC particles (Fig. 1b)

3.2. Mixed adsorption of SHS and TX-100 at AC

Figure 2 presents the adsorption values of SHS on AC surface in the mixed surfactant solutions at different mole fraction of SHS in the solutions. It should be noted that at low total solution concentrations the values of SHS adsorbed from the binary mixtures are higher compare to SHS adsorption from its single solution. It means that SHS adsorption increase in the presence of TX-100. The larger SHS mole fraction in the mixture, the higher SHS adsorption and the larger is the total surfactant concentration at which this effect happen. These findings might be explained by that fact that the embedding of the molecules of the nonionic surfactant in the adsorption SHS layer reduce the electrochemical repulsion between the negatively charged head groups of SHS molecules. This results in higher adsorption of the anionic surfactant from SHS/TX-100 mixture, than that observed in single SHS solution.

In contrast, at higher total surfactants concentration, the amount of SHS adsorbed from the binary mixtures is lower compare to SHS adsorption from its single solution. At higher total concentration the ionic surfactant is possibly displaced by the nonionic on AC surface. The displacement is higher at larger mole fraction of nonionic surfactants in the solution (Figure 2). This means that interraction of the nonionic surfactant to AC surface is stronger compare to that one for SHS.

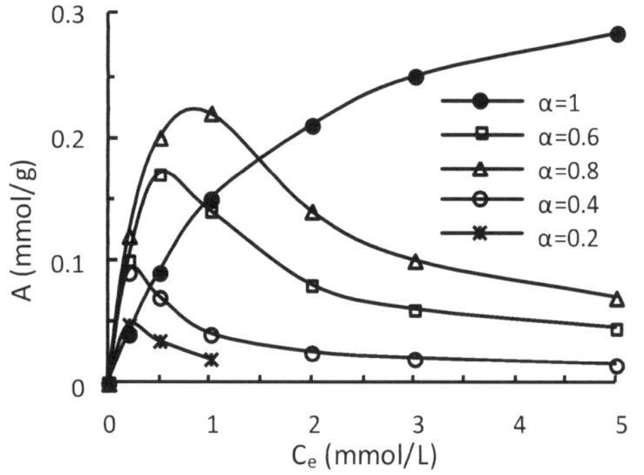

Figure 2. Adsorption of SHS on AC from mixed surfactants solutions at different mole fraction (α) of SHS in the mixture.

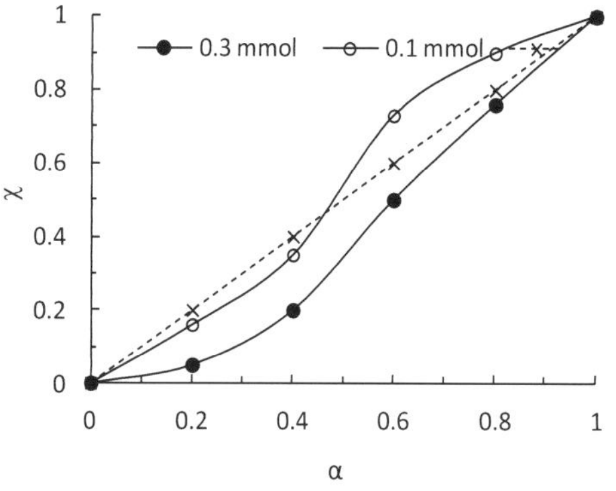

Figure 3. SHS mole fraction on AC surface versus SHS mole fraction in the solution at different total concentration of the surfactants mixture.

Figure 3 shows that, a mole fraction of SHS at AC surface essentially varies at different concentrations of the surfactants mixture. If the surfactant composition on AC surface were the same as that in the bulk solution, the experimental points would fall in the median line in the Figure 3. It seems that at low total concentration of SHS/TX-100 mixture (0.1 mmol/L) and at a high SHS mole fraction in the solution (α> 0.5) the adsorption layer is enriched with SHS molecules. This might be due to smaller electrostatic repulsion between the negatively charged head groups of SHS owing to incorporation of nonionic TX-100 molecules in the mixed adsorbed layer. As was discussed above, an increase of the concentration of nonionic surfactants in the mixture leads to displacement of anionic surfactant from AC surface and this will reduce SDS mole fraction in the mixed adsorption layer. For relatively low total surfactants concentration (0.1 mmol/L) and at the SHS mole fraction in the solution of 0.5, the mixed adsorption layer is formed from equal amounts of both surfactants. At higher surfactant concentration (0.3 mmol/L), SHS adsorbed at AC is more efficiently replaced by TX-100 molecules and as a result the mixed adsorption layer is enriched with the molecules of the nonionic surfactant.

4. Conclusions

Composition of the adsorption layers on AC surface in single and mixed solutions of SHS and TX-100 surfactants have been studied. The presented data indicates that the surfactants adsorption from their single solutions is of Langmuir type. The results show that the surfactants composition at the AC surface differs essentially from the solution composition of the surfactants. At low concentrations of the surfactants mixture, SHS adsorption is higher in the presence of nonionic surfactant compare to SHS adsorption in the single solution. Obviously, incorporation of the molecules of the nonionic surfactant in the mixed adsorption layer will reduce electrochemical repulsion between the charged hydrophilic groups of the anionic surfactant and thus promoting SHS adsorption. At higher total surfactants concentration in the solution, the amount of the adsorbed SHS from the binary mixtures is lower compare to SHS adsorption from its single solutions due to displacement of the ionic surfactant from AC surface by the nonionic counter partner.

Acknowledgement
O. Kochkodan, N. Antraptseva and V. Maksin gratefully acknowledge the support of Ministry of Education and Science of Ukraine (Grant No. 110/103-F).

References
[1] Rosen M J and Kunjappu J M Surfactants and interfacial phenomena New Jersey: Jon Willey and Songs Inc.; 2012, 616 p
[2] Bagheri A and Paresa K 2017 Synergism between non-ionic and cationic surfactants in a concentration range of mixed monolayers at an air–water interface *RCS Advances* **7** 18151-18161
[3] Moussavi G, Shekoohiyan S, and Mojab S 2016 Adsorption capacity of NH4Cl-induced activated carbon for removing sodium dodecyl sulfate from water *Desalination and Water Treatment* **57** 11283–11290
[4] Ying G G 2006 Fate, behavior and effects of surfactants and their degradation products in the environment *Environmental International* **32** 417-431
[5] Sweetman M J, May S, Mebberson N, Pendleton P, Vasilev K, Plush S E and Haybal J D 2017 Activated Carbon, Carbon Nanotubes and Graphene: Materials and Composites for Advanced Water Purification *Journal of Carbon Research* **3** 18-42
[6] Yin C Y, Aroua M K and Daud W M A W 2007 Review of modifications of activated carbon for enhancing contaminant uptakes from aqueous solutions *Separation and Purification Technology* **52** 403-415
[7] Soria-Sánchez M, Maroto-Valiente A, Guerro-Ruiz A and Nevskaia D M 2010 Adsorption of non-ionic surfactants on hydrophobic and hydrophilic carbon surfaces *Journal of Colloid and Interface Science* **343** 194-199.
[8] Marsalek R, Pospisil J and Taraba B 2011 The influence of temperature on the adsorption of CTAB on coals *Colloids and Surfaces A: Physicochemical and Engineering Aspects* **383** 80–85
[9] Kochkodan O D, Klimenko N A and Karmazina T V 1996 Thermodynamic characteristics of adsorption of non-ionic surfactants onto acetylene carbon black and AG-3 activated carbon *Colloid Journal* **58** 330-335
[10] Goncharuk V V, Karmazina T V, Slisenko V I, Vasilkevich A A, Kochkodan O D, Valkovska N I, Ivanitskii P G and Krotenko V T 2005 Effect of adsorption treatment on molecular-dynamic state of water *Journal of Water Chemistry and Technology* **16** 8-13
[11] Xiao J X, Zhang Y, Wang C, Zhang J, Wang C M, Bao Y X and Zhao Z G 2005 Adsorption of cationic–anionic surfactant mixtures on activated carbon *Carbon* **43** 1032–1038
[12] Gao Y, Zhang Y, Huang Z and Gu T 1989 *Acta Scientiarum Naturalium Universitatis Pekinensis* **25** 726-732
[13] Sun J, Du K, Fu L, Gao J, Zhang H, Feng W and Ji P 2014 Sodium Hexadecyl Sulfate as an Interfacial Substance Adjusting the Adsorption of a Protein on Carbon Nanotubes *ACS Applied Materials and Interfaces* **6** 15132-15139.
[14] Julien F, Baudu M and Mazet M 1998 Relationship between chemical and physical surface properties of activated carbon *Water Research* **32** 3414–3424.

Lifetime study for solution processed organic light emitting diodes

Fuh-Shyang Juang[1,a], Chang-Chi Lee[1,b], Jeng-Yue Chen[2,c], Yen-Hua Lin[1,d] and Ding-Wen Zhang[1,e]

[1]Institute of Electro-Optical and Materials Science, National Formosa University, Huwei, Yunlin, Taiwan
[2]Department of Electrical Engineering, National Formosa University, Huwei, Yunlin, Taiwan

[a]fsjuang@nfu.edu.tw; [b]micro@maxcable.com; [c]buo@nfu.edu.tw; [d]selb7413@gmail.com; [e]xp84050816@gmail.com

Abstract. In this paper, blue fluorescent organic light-emitting diodes (OLED) are fabricated by solution processes. UV Ozone exposure time on the ITO surface and the thickness of the hole injection layer (HIL), electron transport layer (ETL) and electron injection layer (EIL) are modulated to study the effects on device characteristics and lifetime. ITO/MADN:13% UBD-07/Al is used as the basic structure. First, HIL (PEDOT: PSS) is added. It is found that although the hole current is increased, the device does not emit light. Secondly, ETL (TPBi) is added to facilitate the injection of electrons from cathode to EML, and in such a way, hole-electron can be recombined in EML and emit light. With an increase in the time to expose the ITO into UV Ozone from 100s and 200s to 400s, the total current of the OLED is increased, and luminance is improved, but the current efficiency is not obviously improved. Its role facilitating the electron injection is observed by adjusting the LiF thickness of EIL. In addition to increasing the hole current, the addition of PEDOT as HIL may reduce the chance of a device shorten resulting from the spikes on ITO surface, and the lifetime of the OLED with this structure is 121 min. In order to prolong device lifetime and better balance the injection quantity of electrons and holes, the structure of ETL is adjusted to be TPBi+Alq3. When TPBi+Alq3 is 10nm/15nm thick, current efficiency is 1.29 cd/A at 8V, and the lifetime is extended to 252 min.

Keywords: organic light emitting diode; solution process; blue fluorescent; hole injection layer; electron transport layer; lifetime

1. Introduction

The blue organic light-emitting diode (blue OLED) is used for full color flat-panel display and lighting. The blue emitter is used as one of three main color sources, and compared with green and red OLED, traditional blue OLED has disadvantages such as low efficiency and a short lifetime [1]. Once blue light attenuates, it is possible to cause a color shift on the whole screen, and all manufacturers are forced to study a solution to this disadvantage. In addition to a short lifetime, OLED has other

disadvantages such as the deterioration of its organic materials resulting from water oxidation and oxygen. How to prolong the device's lifetime is a very important research subject.
K.F. Jeltsch et al [2] fabricated OLEDs by solution process and studied the aggregation of Flrpic on PEDOT: PSS to prolong the lifetime of the OLEDs. Device structure is ITO/PEDOT: PSS/PVK: FIrpic: OXD-7/Ca/Al. During the research of blue phosphorescence OLED, Sang Kyu Jeon et al [3] inserted a host material between HTL and EML as an exciton blocking layer to prolong the lifetime by confining the recombination zone in EML. Device structure is ITO/DNTPD/BPBPA/ mCBP/mCBP: Ir(dbi)3/LG201/LiF/Al. Jang Hyuk Kwon et al [4] used Liq as EIL material for OLED and found that double electron injection layers improved current efficiency by 20% and prolonged the lifetime by 40%. Device structure is ITO/NPD/EML/Alq3:Liq/Liq/Al.

2. Experiments

Table 1. The detailed device parameters (unit: nm)

No	Anode	HIL	EML	HBL	ETL	EIL	Cathode
	ITO UV O3	PEDOT (spin)	MADN : 13%UBD-07 (spin)	TPBi (evap.)	Alq_3 (evap.)	LiF (evap.)	Al (evap.)
A-1	400s	55	45	0	0	0	150
A-2	400s	0	45	0	0	0.8	150
B-1	400s	0	45	0	0	0.8	150
B-2	400s	0	45	20	0	0.8	150
C-1	100s	0	45	20	0	0.8	150
C-2	200s	0	45	20	0	0.8	150
C-3	400s	0	45	20	0	0.8	150
D-1	100s	0	45	20	0	0.5	150
D-2	100s	0	45	20	0	0.8	150
E-1	100s	55	45	20	0	0.5	150
E-2	100s	55	45	10	15	0.5	150

The emitting layer solution was prepared in advance by mixing host material MAND: dopant UBD-07 = 8.7mg: 1.3mg, and dissolved in 1 cc chlorobenzene. In N_2 glove box, 1 cc PEDOT: PSS was filtered by 0.45 μm filter, spin coated onto the ITO glass substrate at 2500 rpm for 30s, and then dried on the hot plate at 110 °C for 20min. The pre-prepared emitting layer solution was filtered by 0.2 μm filter, spin coated on PEDOT: PSS surface at 2500 rpm for 30min, and then dried on the hot plate at 110 °C for 20min [5]. After the spin-coating of EML, TPBi film was deposited by thermal evaporation, then LiF and Al also by thermal evaporation. Finally the device was sent to the glove box for packaging. The encapsulation glass was attached onto the glass substrate by UV cure adhesive, and exposed to UV-light for 170s to cure the UV adhesive. The current-voltage-luminance (I-V-L) measurement is performed by Keithley 2400 (power supply) and Photo Research PR 650 Spectra colorimeter. The L-V of OLED over time driven by a constant current source were measured by a self-designed measurement system (with photodetector, Arduino and Raspberry Pi) for lifetime measurement.

3. Results and discussion

3.1. The effects of hole injection layer (HIL)

In this section, ITO/EML (45 nm)/Al (150 nm) is used as the basic structure to explore the amount of hole injection current. The HIL is added in the device A-1, but not in the device A-2. The band diagrams are shown in Fig.1 and the structural parameters of the device are shown in Table 1. According to the luminance vs. current density shown in Fig.2, it is found that both devices A-1 and A-2 do not emit the light. Because there is about 1.7eV energy barrier between Al cathode and MADN, it may be concluded that the electrons are not injected into EML from the cathodes of A-1 and A-2. According to the I-V characteristics of device A-1 as shown in Fig.2, it can be seen that A-1 is similar to ohmic characteristics. After the insertion of the PEDOT: PSS (HIL), the hole current is increased. Due to no HIL in device A-2, during hole injection, it is required to overcome a higher energy barrier, so the cut-in voltage is higher, about 1V.

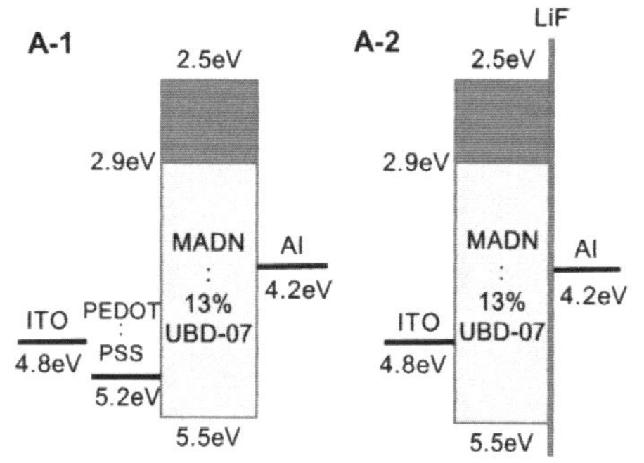

Figure 1. (a) Device A-1, (b) Device A-2 energy band diagram

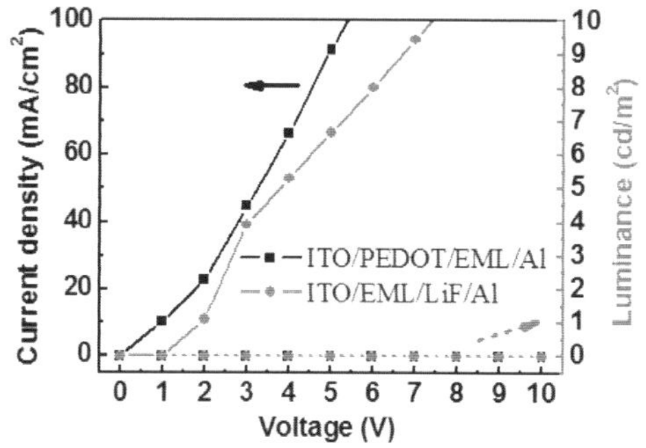

Figure 2. Current density-voltage- luminance curves of device A-1 and A-2.

3.2. The insertion of hole blocking layer (HBL)

According to the results shown in Fig.2 that the device A-2 fails to emit the light, it may be deduced that the addition of LiF does not facilitate cathode electron injection, so ETL TPBi is inserted in the device B-2 to facilitate the electron injection. The device parameters are shown in Table 1, and the device band diagram is shown in Fig.3. The electrons injected amount is increased by the electron transporting characteristic of TPBi. The holes are confined in the emitting layer by the barrier formed from the HOMO energy level difference between EML and HBL (0.7eV) to improve the ability of electron-hole recombination opportunity to emit more light. When TPBi 20nm is inserted, the total current is decreased due to the increase in total resistance, as shown in Fig.4 (a). HBL blocks the holes, causing the hole current to decrease, but ETL TPBi 20 nm facilitates electron injection. In such a way, the recombined hole-electrons are increased to improve the luminance and efficiency of the device, as shown in Fig. 4(b). At 8V, the luminance is 221.2 cd/m^2, and the efficiency is 0.52 cd/A.

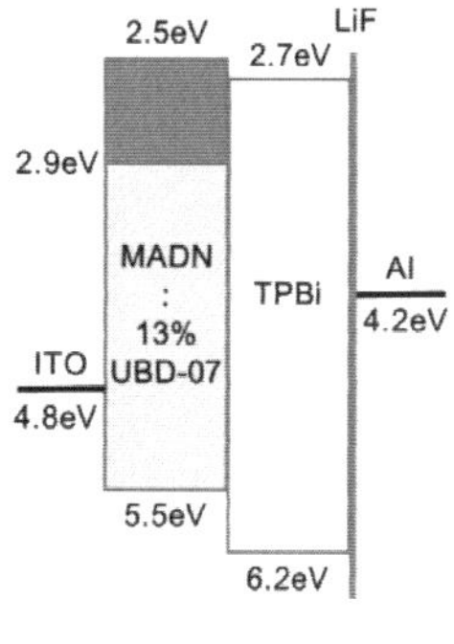

Figure 3. Device B-2 energy band diagram

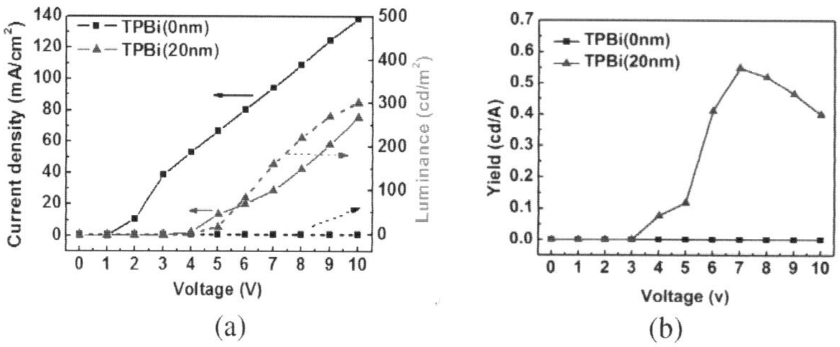

Figure 4. (a) Current density-voltage-luminance curves and (b) Current efficiency-voltage curves of device B-1 and B-2.

3.3. UV Ozone exposure time for ITO

Figure 3 shows the energy band diagram of C-1, C-2 and C-3. In this section, the time of ITO exposed to the UV Ozone is adjusted to compare the hole injection amount. With an increase in the time of exposure to the UV Ozone from 100s, 200s to 400s, the total current of the OLED is increased with the increase of ITO work function by increasing the exposure time to UV Ozone., and luminance is improved, too, as shown in Fig.5(a). But the current efficiency is not obviously improved, as shown in Fig.5(b). Furthermore, with the increase of ITO work function, the ITO-HIL-EML shared bias from applied voltage is reduced, while a partial voltage is transferred to EML-ETL-LiF-Al side which resulting in injection electrons from cathode increased. Therefore, with the increase in the ITO exposure time to UV Ozone, both the electron and hole number is increased simultaneously (and the luminance is also improved simultaneously), but the current efficiency is not obviously improved.

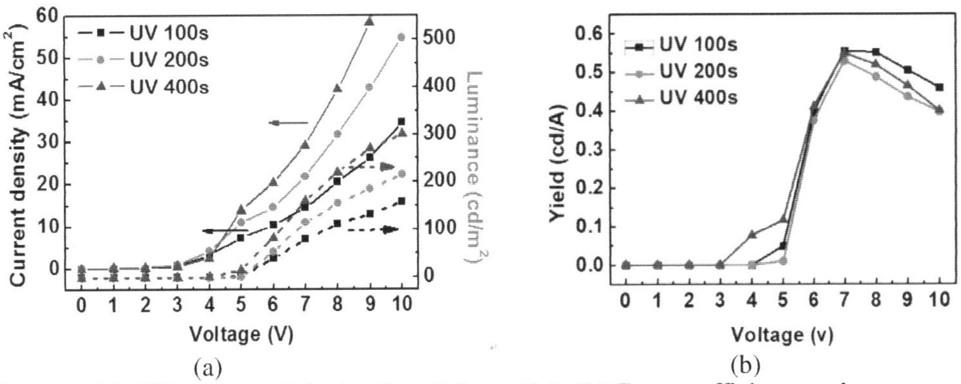

Figure 5. (a) I-V-L curves of device C-1, C-2 and C-3. (b) Current efficiency-voltage curves

3.4. Adjusting the thickness of the electron injection layer (EIL, LiF)

LiF is a non-conductive material [6]. If it is too thick, the resistance of the OLED will increase, and total current will decrease. When the thickness of LiF decreases from 0.8nm to 0.5nm, the electron injection current increases, the cut-in voltage also decreases, and the luminescence and efficiency of LiF 0.5nm are better than that of LiF 0.8nm, as shown in Fig.6.

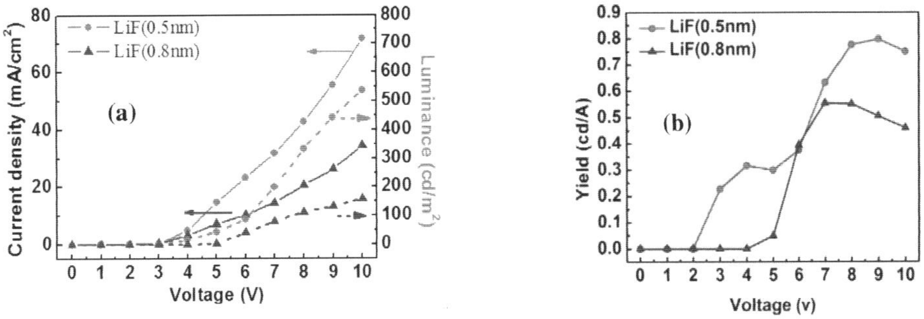

Figure 6. (a) I-V-L curves of device D-1 and D-2. (b) Current efficiency-voltage curves.

3.5. The effects of ETL

As an energy barrier 1.5eV between the work function of Al and the LUMO of TPBi is unfavorable to the electron injection, Alq3 is inserted between TPBi and Al in the device E-2 to slightly reduce the energy barrier for electrons injected from cathode to EML (1.5eV reduced to 1.0eV). If the ETL total thickness of TPBi+Alq3 increases too much, the total resistance of the device may increase additionally which will decrease the device current. So the structure of the device E-1 is used for the device E-2, but the TPBi 20 nm is changed to TPBi/Alq_3 10nm/15nm. Because in device E-2 the energy barrier for electron injection from cathode to EML increases in multi-steps (along the LUMO energy level of Alq3 and TPBi) gradually, as shown in Fig.7, more electrons could inject from cathode to EML. So the device E-2 with ETL as TPBi/Alq3 has a higher current density than device E-1, as shown in Fig.8 (a). The electron-hole recombination ratio in device E-2 is improved to enhance the luminance and efficiency. The efficiency is improved from 0.64 cd/A to 1.29 cd/A, as shown in Fig. 8 (b). As the electron injection current is relatively balanced with the hole injection current, the half-decay lifetime is prolonged from 121min to 252.2min, as shown in Fig.9. Regarding the comparison of electron mobility, the electron mobility of TPBi is about 3.3-8.0 $x10^{-5}$ cm^2/V which is higher than Alq3 0.4-1.1$x10^{-5}$ cm^2/Vs. However, Fig. 8(a) shows that the device E-1 has lower current density than E-2, so it is deducted that after the insert of Alq3 as part of ETL, the device lifetime is prolonged mainly because LUMO energy level of Alq3 can facilitate electron injection from cathode and reduce electron accumulation on cathode.

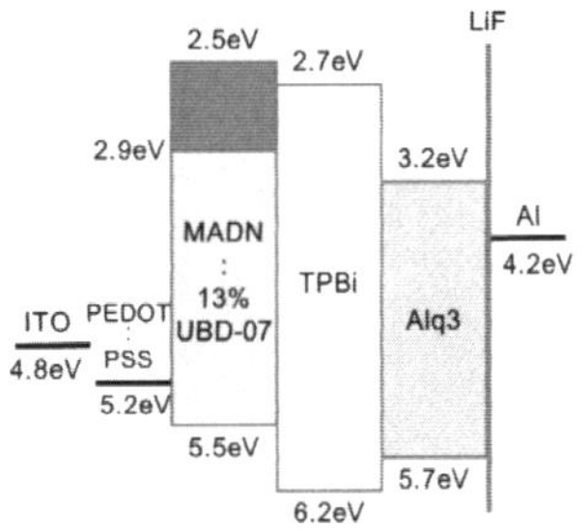

Figure 7. Device E-2 energy band diagram.

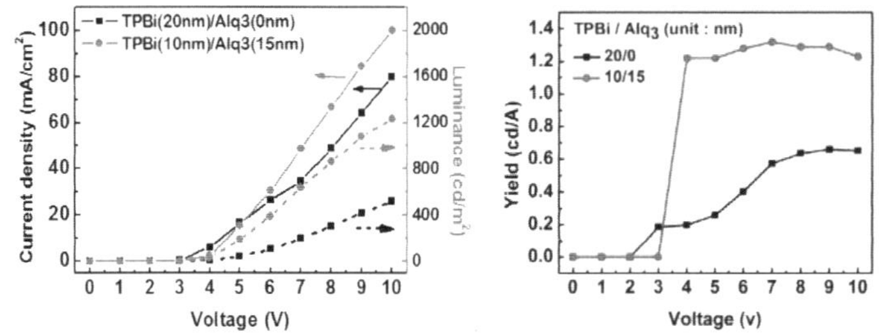

Figure 8. (a) Current density-voltage-luminance curves and (b) Current efficiency- voltage curves of device E-1 and E-2.

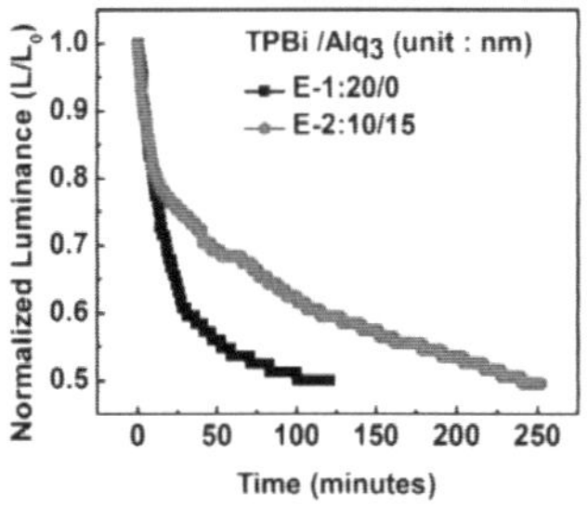

Figure 9. Lifetime comparisons of device E-1 with E-2

4. Summary

The HIL (PEDOT: PSS) is added first, and it is found that although the hole current is increased, the device does not emit light. The ETL (TPBi) is added to facilitate the electron injection from cathode to EML, and in such way, the hole-electron can recombine in EML and emit light. At 8V, the luminance is 221.2 cd/m^2, and the efficiency is 0.52 cd/A. By adjusting UV Ozone exposure time, the hole injection current could be controlled. When LiF thickness is reduced from 0.8 to 0.5nm, the electron injection current increases, the cut-in voltage decreases, the luminance and the efficiency (0.78 cd/A@8V) are improved. In addition to increasing the hole current, the insertion of PEDOT: PSS as HIL can reduce the chance of short circuit of the device resulting from a spike effect on the ITO surface. The lifetime of the device ITO/PEDOT: PSS/EML/TPBi/LiF/Al is 121 min. In order to better balance the electron and hole injection amount so as to prolong the device's lifetime, the ETL structure is adjusted to be TPBi+Alq3. When TPBi/Alq3 is 10nm/15nm thick, current efficiency increases to 1.29 cd/A at 8V, and the lifetime is prolonged to 252 min. In the future, the thickness of HIL and EML will be adjusted and tried to extend the lifetime of OLEDs further more.

References

[1] Chong Li, Taiju Tsuboi, and Wei Huang, Optical Materials, Vol. 36 (2014), p. 1720.
[2] K. F. Jeltsch, G. Lupa, R. T. Weitz, Organic Electronics, Vol. 26 (2015), p. 365.
[3] Jeong-A Seo, Sang Kyu Jeon, Myoung Seon Gong, Jun Yeob Lee, Chang Ho Noh and Sung Han Kim, J. Mater. Chem. C, Vol. 3 (2015), p. 4640.
[4] Jang Hyuk Kwon, Jun Yeob Lee, Synthetic Metals, Vol. 159 (2009), p.1292.
[5] D. K. Dubey, M. Singh, S. Sahoom, and J. H. Jou, Microelectronics Reliability, Vol. 83 (2018), p. 293.
[6] L. S. Hung, C.W. Tang, and M. G. Mason, Appl. Phys. Lett., Vol. 70 (1997), p. 152.

IOP Conf. Series: Materials Science and Engineering **600** (2019) 012019 doi:10.1088/1757-899X/600/1/012019

SBFE analysis of nano-scale elastic layer with consideration of surface energy effects

Chantha Chhuon[1, a], **Jaroon Rungamornrat**[1, b] **and Sawekchai Tangaramvong**[1, c]

[1]Applied Mechanics and Structures Research Unit, Department of Civil Engineering, Faculty of Engineering, Chulalongkorn University, Bangkok 10330, Thailand

[a]chhuonchantha09@gmail.com; [b]jaroon.r@chula.ac.th; [c]Sawekchai.t@chula.ac.th

Abstract. This paper presents an efficient and accurate numerical technique for determining the mechanical response of an infinite elastic layer under surface loading and surface stress effects. The governing equation of the bulk is formulated from the classical linear elasticity theory via a SBFE technique whereas that of the material surface is obtained from a complete version of Gurtin-Murdoch surface elasticity theory. By enforcing the continuity at the interface of the material surface and the bulk, it leads to a system of linear non-homogenous ordinary differential equations governing the nodal functions. A general solution of the resulting system of ODEs is constructed via standard procedures and then used together with the boundary conditions to form a system of linear algebraic equations governing nodal degrees of freedom. To investigate the accuracy and convergence of the proposed method, selected scenarios are solved and obtained numerical results are reported and discussed.

1. Introduction

In the early of 21st century, a nano-layer surface coating has been commonly utilized to enhance both surface and overall properties of components such as the energy harvested reflective color filter, surface wear resistance complex metal boron-carbide, and invisible carbon nanotube coating. Although experimental studies and theoretical simulations based on discrete models [1, 2] provide a profound understanding reflecting the real response of surface coatings, they result in the prohibitive cost and large computational effort, respectively. Modified continuum-based models, with the incorporation of Gurtin-Murdoch surface elasticity theory [3, 4]with the classical linear elasticity, are considered promising in that they can provide the first approximation of the response prediction of nano-size problems requiring relatively cheap computational effort.

In the past studies, the complete version of Gurtin-Murdoch model has been widely adopted to investigate the surface energy effect of various nano-size problems under prescribed loading by various analytical solution methods such as Fourier integral transform for buried load in an infinite elastic layer [5]; Hankel integral transform with Love's representation for an infinite rigid-based elastic layer under axisymmetric surface load [6] and a layered elastic half-space under axisymmetric surface load [7]; and a technique based on Boussinesq potential for an elastic half-space subject to surface circular shear traction [8]. Although those employed solution procedures are successful in predicting the response of nano-size problems via the Gurtin-Murdoch model such as the size-dependency behavior, the issue of computational efficiency arises when more practical and large scale problems become of interest.

The present study aims to implement an alternative solution procedure, based upon the framework of a scaled boundary finite element method (SBFEM), to obtain response of nano-size problems via a continuum model enhanced by integrating Gurtin-Murdoch surface elasticity theory to treat the influence of surface energy. In particular, a problem associated with two-dimensional, infinite, rigid-based, elastic layer under surface loading is chosen as the representative case to demonstrate the capability of the proposed technique.

2. Problem formulation

Consider a two-dimensional, infinite, elastic layer of thickness h resting on a rigid foundation and subjected to arbitrarily distributed normal traction p over the length $2a$ as shown schematically in Fig. 1. The bulk of the layer is made of a homogeneous, isotropic, linearly elastic material whereas a thin layer on the top of the layer (termed here the surface) is made of a different homogeneous, isotropic, linear elastic material with the presence of initial residual surface tension. In the present study, the layer is assumed to be free of body force.

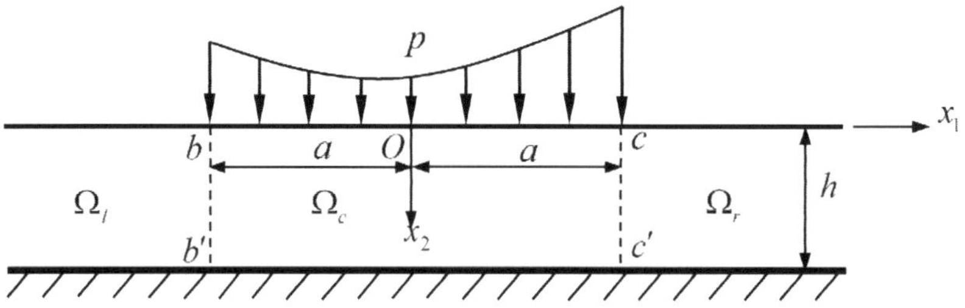

Figure 1. Two-dimensional, elastic layer subjected to arbitrary vertical pressure

The elastic layer can be divided into three sub-domains along the lines bb' and cc' and they are referred to the left part Ω_l, the center part Ω_c, and the right part Ω_r. In the formulation presented further below, a representative domain $\hat{\Omega}$ with a similar shape as that of the center part Ω_c is considered. The left, right, top and bottom boundaries of this representative domain $\hat{\Omega}$ are denoted by $\partial\hat{\Omega}^l$, $\partial\hat{\Omega}^r$, $\partial\hat{\Omega}^t$, and $\partial\hat{\Omega}^b$, respectively. Once the formulation is achieved on $\hat{\Omega}$, it can be readily applied to all subdomains Ω_l, Ω_c, and Ω_r. For instance, Ω_l is obtained from $\hat{\Omega}$ by taking the boundary $\partial\hat{\Omega}^l$ to infinity and removing all loading on the boundary $\partial\hat{\Omega}^t$.

From the classical theory of elasticity, the equilibrium equations, the constitutive laws, and the infinitesimal strain-displacement relationship of the bulk material can be expressed in a concise form, for a two-dimensional body subjected to plain-strain condition and zero body force, as

$$\boldsymbol{L}^T\boldsymbol{\sigma}^b = \boldsymbol{0} \;;\; \boldsymbol{\sigma}^b = \boldsymbol{D}\boldsymbol{\varepsilon}^b \;;\; \boldsymbol{\varepsilon}^b = \boldsymbol{L}\boldsymbol{u}^b \tag{1}$$

where $\boldsymbol{\sigma}^b, \boldsymbol{\varepsilon}^b, \boldsymbol{u}^b, \boldsymbol{D}$ and $\boldsymbol{L} = \boldsymbol{b}_1\partial/\partial x_1 + \boldsymbol{b}_2\partial/\partial x_2$ are a vector containing independent in-plane stress components, a vector containing independent in-plane strain components, a vector containing in-plane displacement components, a modulus matrix involving material constants, and a two-dimensional, linear differential operator with $\boldsymbol{b}_1$ and $\boldsymbol{b}_2$ denoting Boolean matrices, respectively; the superscript " T " denotes the matrix transpose operator; and the superscript " b " is used to emphasize field quantities associated with the bulk material. By applying the standard weighted residual technique together with the integration by parts to the basic governing equations (i.e., Eq. (1)), the alternative weak statement governing the bulk material can be expressed as:

$$\int_{\hat{\Omega}} (\boldsymbol{L}\boldsymbol{w})^T \boldsymbol{D}(\boldsymbol{L}\boldsymbol{u}^b)dA = \int_{\partial\hat{\Omega}} \boldsymbol{w}^T \boldsymbol{t}^{b(\partial\hat{\Omega})} ds \tag{2}$$

where $\boldsymbol{w}$ denotes a sufficiently smooth weighted function; $\partial\hat{\Omega}=\partial\hat{\Omega}^{l}\cup\partial\hat{\Omega}^{r}\cup\partial\hat{\Omega}^{t}\cup\partial\hat{\Omega}^{b}$; and $\boldsymbol{t}^{b(\partial\hat{\Omega})}$ is the traction acting to the boundary $\partial\hat{\Omega}$ of the bulk. Now, let one introduce the scaled boundary finite element approximation of the displacement $\boldsymbol{u}^{b}$ and weighted function $\boldsymbol{w}$ such that

$$\boldsymbol{u}^{b}=\boldsymbol{u}^{b}(x_1,x_2)=\sum_{i=1}^{p}\phi^{i}(x_2)u^{i}(x_1)=\boldsymbol{NU}\;;\;\boldsymbol{w}=\boldsymbol{w}(x_1,x_2)=\sum_{i=1}^{p}\phi^{i}(x_2)w^{i}(x_1)=\boldsymbol{NW} \tag{3}$$

where $\boldsymbol{N}$, $\boldsymbol{U}$, and $\boldsymbol{W}$ represent, respectively, a matrix containing the basic functions ϕ^{i}, a vector containing the nodal displacement functions u^{i}, and a vector containing the nodal weighted functions w^{i}. By substituting the approximations of $\boldsymbol{u}^{b}$ and $\boldsymbol{w}$ into Eq.(2) together with the integration by parts of terms involving the first derivative of the weighted function, and by invoking the arbitrariness of the vector $\boldsymbol{W}$, we obtain the scaled boundary finite element equation of the bulk material and the corresponding boundary conditions as follows:

$$\boldsymbol{E}_2^{b}\boldsymbol{U}''+[\boldsymbol{E}_1^{b}-(\boldsymbol{E}_1^{b})^{T}]\boldsymbol{U}'-\boldsymbol{E}_0^{b}\boldsymbol{U}+\boldsymbol{T}^{bt}+\boldsymbol{T}^{bb}=\boldsymbol{0}\;;\boldsymbol{Q}^{bl}=\boldsymbol{Q}^{b}(x_{1l})=-\boldsymbol{P}^{bl}\;;\boldsymbol{Q}^{br}=\boldsymbol{Q}^{b}(x_{1r})=\boldsymbol{P}^{br} \tag{4}$$

where $\boldsymbol{T}^{bt}=\boldsymbol{N}^{T}(0)\boldsymbol{t}^{bt}(x_1)$, $\boldsymbol{T}^{bb}=\boldsymbol{N}^{T}(h)\boldsymbol{t}^{bb}(x_1)$, $\boldsymbol{P}^{bl}=\int_0^h\boldsymbol{N}^{T}\boldsymbol{t}^{bl}dx_2$, $\boldsymbol{P}^{br}=\int_0^h\boldsymbol{N}^{T}\boldsymbol{t}^{br}dx_2$, $\boldsymbol{B}=\dfrac{\partial\boldsymbol{N}}{\partial x_1}$

$$\boldsymbol{E}_2^{b}=\int_0^h\boldsymbol{B}^{T}\boldsymbol{b}_2^{T}\boldsymbol{D}\boldsymbol{b}_2\boldsymbol{B}dx_2,\;\boldsymbol{E}_1^{b}=\int_0^h\boldsymbol{N}^{T}\boldsymbol{b}_1^{T}\boldsymbol{D}\boldsymbol{b}_2\boldsymbol{B}dx_2,\;\boldsymbol{E}_0^{b}=\int_0^h\boldsymbol{N}^{T}\boldsymbol{b}_1^{T}\boldsymbol{D}\boldsymbol{b}_1\boldsymbol{N}dx_2,$$

$\boldsymbol{t}^{bt},\boldsymbol{t}^{bb},\boldsymbol{t}^{bl},\boldsymbol{t}^{br}$ are tractions on the boundaries $\partial\hat{\Omega}^{t}$, $\partial\hat{\Omega}^{b}$, $\partial\hat{\Omega}^{l}$, and $\partial\hat{\Omega}^{r}$, respectively, $x_1=x_{1l},x_1=x_{1r}$ denote the coordinates associated with the left and right boundaries, respectively, and $\boldsymbol{Q}^{b}$ is termed the internal nodal flux of the bulk material defined by

$$\boldsymbol{Q}^{b}=\boldsymbol{E}_1^{b}\boldsymbol{U}+\boldsymbol{E}_2^{b}\boldsymbol{U}' \tag{5}$$

For the material surface, its response is modeled by the complete version of Gurtin-Murdoch surface elasticity theory [3, 4]. In particular, the in-plane and out-of-plane equilibrium equations of the surface in terms of the surface displacements, for the one-dimensional case, and the corresponding boundary conditions are given explicitly by

$$\boldsymbol{E}^{s}\boldsymbol{U}^{s''}+\boldsymbol{t}^{0}+\boldsymbol{t}^{sb}=\boldsymbol{0}\;;\boldsymbol{Q}^{sl}=\boldsymbol{Q}^{s}(x_{1l})=-\boldsymbol{t}^{sl}\;;\boldsymbol{Q}^{sr}=\boldsymbol{Q}^{s}(x_{1r})-\boldsymbol{t}^{sr} \tag{6}$$

where the superscript "s" is used to designate quantities associated with the surface; $\boldsymbol{t}^{0}$ is the prescribed loading on the surface; $\boldsymbol{t}^{sb}$ denotes the traction exerted to the surface by the bulk; $\boldsymbol{t}^{sl}$, $\boldsymbol{t}^{sr}$ are tractions on the left and right boundaries of the surface;

$$\boldsymbol{E}^{s}=\begin{bmatrix}(2\mu^{s}+\lambda^{s}) & 0\\ 0 & \tau^{s}\end{bmatrix},\;\boldsymbol{U}^{s}=\begin{Bmatrix}u_1^{s}\\ u_2^{s}\end{Bmatrix} \tag{7}$$

in which μ^{s} and λ^{s} are surface Lame constants, τ^{s} is the residual surface tension, and u_1^{s}, u_2^{s} are components of the surface displacement; and, similar to the bulk material, $\boldsymbol{Q}^{s}$ denotes the internal flux within the material surface defined by $\boldsymbol{Q}^{s}=\boldsymbol{E}^{s}\boldsymbol{U}^{s'}$. By enforcing the continuity along the interface between the bulk and the material surface, it leads to the final system of governing differential equations of the whole system and the corresponding total nodal internal flux

$$\boldsymbol{E}_2^{bs}\boldsymbol{U}''+[\boldsymbol{E}_1^{b}-(\boldsymbol{E}_1^{b})^{T}]\boldsymbol{U}'-\boldsymbol{E}_0^{b}\boldsymbol{U}=\boldsymbol{t}\;;\;\boldsymbol{Q}=\boldsymbol{E}_1^{b}\boldsymbol{U}+\boldsymbol{E}_2^{bs}\boldsymbol{U}' \tag{8}$$

where $\boldsymbol{E}_2^{bs}$ is a matrix resulting from the assembly of $\boldsymbol{E}_2^{b}$ and $\boldsymbol{E}^{s}$, $\boldsymbol{t}$ denotes a vector containing the information of the side-face tractions, and $\boldsymbol{Q}$ denotes the total nodal internal flux. By further enforcing the prescribed displacement at the bottom of the layer, the nodal displacement vector $\boldsymbol{U}$ and nodal traction vector $\boldsymbol{t}$ can be partitioned into $\boldsymbol{U} = \{\boldsymbol{U}^f \quad \boldsymbol{U}^r\}$ and $\boldsymbol{t} = \{\boldsymbol{t}^f \quad \boldsymbol{t}^r\}$ where $\boldsymbol{U}^f, \boldsymbol{t}^r$ are unknown a priori and $\boldsymbol{U}^r, \boldsymbol{t}^f$ are fully prescribed. According to such partition, the system (8) can also be portioned to arrive at the reduced system of differential equations governing the unknown displacement $\boldsymbol{U}^f$.

3. Solution methodology

Following the standard procedure in the theory of differential equations, the general solution of the unknown displacement $\boldsymbol{U}^f$ can be readily established. In particular, the homogeneous solution, denoted by $\boldsymbol{U}^{fh}$, is obtained by the technique of assuming a solution form and this finally leads to solving a linear eigen problem. A selected efficient algorithm is adopted to accurately and efficiently determine all the eigenvalues and eigenvectors. To construct the particular solution of the unknown displacement $\boldsymbol{U}^{fp}$ for the prescribed traction on the top of the layer (which is assumed expressible in a polynomial form), a well-known method of undetermined coefficients is applied.

By using the general solution of the unknown displacement $\boldsymbol{U}^f$ together with the boundary conditions at the left and right boundaries of the layer, it results in a system of linear algebraic equations governing the unknown data on the boundaries of the layer. Such results for the representative domain can then be used to generate three sub-systems of linear equations governing the unknown data on the boundary of the left part Ω_l, the center part Ω_c, and the right part Ω_r. Finally, by enforcing the continuity of the displacement and traction along the interface between the left part and the center part (i.e., the line bb') and the interface between the right part and the center part (i.e., the line cc') and the standard assembly procedure, it yields a final system of linear algebraic equations governing the nodal displacement data along the two interfaces bb' and cc'. Once such primary unknowns are solved, the elastic field including the displacement and stress within the layer can be readily post-processed with the use of basic field equations (1) and the approximations (3).

4. Preliminary results

To verify the proposed technique, a representative case associated with an elastic half-plane subjected to a constant pressure p over the length $2a$ and without the surface stress effect is considered since the exact solution of the elastic field is available for the comparison purpose. In the numerical study, three levels of discretization across the thickness of the layer (associated with 4, 8, and 16 quadratic elements) are adopted to confirm the convergence of computed numerical solutions and the ratio $h/a = 200$ is chosen to simulate the half-plane case. Results of non-zero stress components along the x_2-axis are reported in Fig. 2(a) together with the exact solution. It is seen that the convergence of the computed numerical solutions and the excellent agreement with the reference solution are observed.

To further demonstrate the capability of the proposed technique to handle nano-scale problems with the surface stress effects, consider an elastic layer of the thickness $h = 8\Lambda$ (where Λ is an intrinsic length scale of the material defined in [5, 6]) under the constant pressure p over the length $a = 2\Lambda$. The material properties for the bulk and the surface used in the numerical study are taken from [6]. Converged results of the stress components along the x_2-axis obtained from the proposed technique are reported in Fig. 2(b) for the two cases (with and without surface stresses). It is apparent that the presence of the surface stresses can significantly alter the elastic response of the layer when the characteristic length of the problem is comparable to the material length scale Λ.

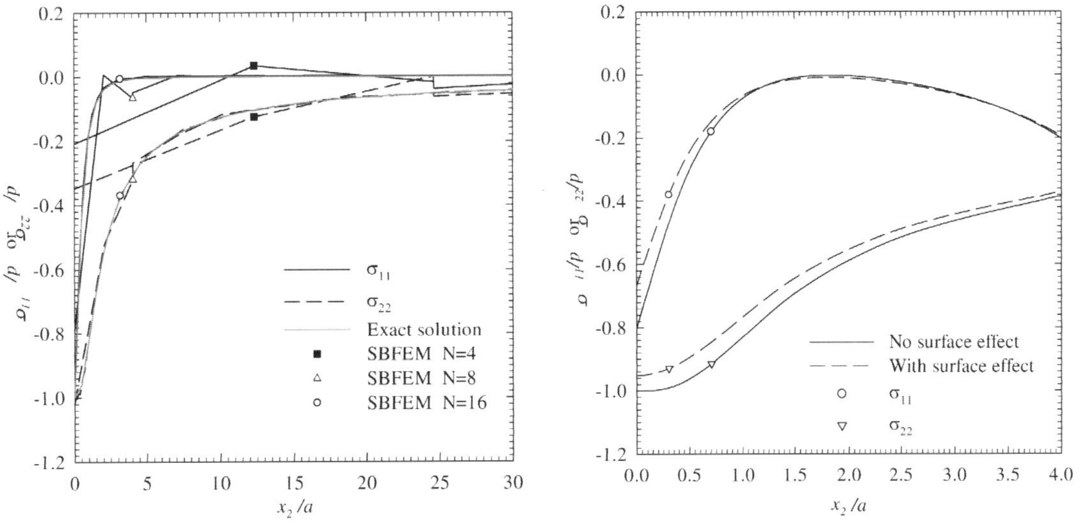

Figure 2. Non-dimensional stress field of (a) elastic half-plane and (b) elastic layer

5. Conclusion

An efficient and accurate numerical technique based on SBFEM has been successfully implemented to determine mechanical response of an elastic layer under the surface loading and surface stress effects. Based on results from a preliminary numerical study, the proposed technique has been found promising and computationally robust. The proposed technique will be used to solve more complicated problems including nano multilayer media and functionally graded material.

Acknowledgments

The authors gratefully acknowledge the support provided by the JICA Project for AUN/SEED-Net.

References

[1] Fereidoon, A., et al., *Atomistic simulation study of mechanical properties of periodic graphene nanobuds.* Computational Materials Science, 2015. **107**: p. 163-169.
[2] Zhang, L., et al., *Atomistic simulation of Al-graphene thin film growth on polycrystalline Al substrate.* Applied Surface Science, 2018. **433**: p. 540-545.
[3] Gurtin, M.E. and A.I. Murdoch, *A continuum theory of elastic material surfaces.* Archive for Rational Mechanics and Analysis, 1975. **57**(4): p. 291-323.
[4] Gurtin, M.E., J. Weissmüller, and F. Larché, *A general theory of curved deformable interfaces in solids at equilibrium.* Philosophical Magazine A, 1998. **78**(5): p. 1093-1109.
[5] Intarit, P., et al., *Surface elasticity and residual stress effect on the elastic field of a nanoscale elastic layer.* Interaction and Multiscale Mechanics, 2011. **4**: p. 85-105.
[6] Rungamornrat, J., P. Tuttipongsawat, and T. Senjuntichai, *Elastic layer under axisymmetric surface loads and influence of surface stresses.* Applied Mathematical Modelling, 2016. **40**(2): p. 1532-1553.
[7] Tirapat, S., T. Senjuntichai, and J. Rungamornrat, *Influence of Surface Energy Effects on Elastic Fields of a Layered Elastic Medium under Surface Loading.* Advances in Materials Science and Engineering, 2017. **2017**: p. 11.
[8] Mi, C., *Surface mechanics induced stress disturbances in an elastic half-space subjected to tangential surface loads.* European Journal of Mechanics - A/Solids, 2017. **65**: p. 59-69.

IOP Conf. Series: Materials Science and Engineering **600** (2019) 012020 doi:10.1088/1757-899X/600/1/012020

Waste from renewable sources as a polyethylene filler

Janusz W. Sikora, Karolina Głogowska, Łukasz Majewski

Lublin University of Technology, Mechanical Engineering Faculty, Department of Technology and Polymer Processing

janusz.sikora@pollub.pl

Abstract. The paper presents research on the impact of three organic fillers from renewable sources, i.e. ground wheat bran, pumpkin seed husks and peanut shells, constituting production waste of the food industry, on selected mechanical and thermal properties of composite injection moldings based on LDPE. The basic strength properties were tested in a static tensile test, hardness and heat deflection temperature, Vicat softening point, and the microscopic structure of the obtained polymer compositions was analysed. The tests were carried out with a different mass content of fillers of 0, 5, 10, 15 and 20% and variable size of filler grains. Samples for testing were obtained in the injection molding process. An increase in the Young's modulus and hardness as well as a decrease in the remaining tested strength properties along with an increase in the filler's share content were found.

1. Introduction

Promoting the use of natural waste in various industries, including polymer processing, is associated with the pressure of global organizations responsible for environment protection, on ecological issues, such as recycling, landfill and waste disposal. Interdisciplinary approach to the subject of polymer processing makes it possible to use materials that until now had no practical application. The increase in the use of natural waste in polymer processing can contribute to reducing the negative impact of this industry on the environment, by accelerating the degradation of polymer components and reducing the consumption of petrochemical polymers.

The properties of filled materials depend not only on the type and geometrical properties of the filler, processed polymer, the presence of additional components, the constructional solution of the plasticizing system, processing conditions [1], but also on the filler content and particle size [2, 3, 4]. For example, smaller particles bond more easily and form a more durable structure [5, 6]. The combination of wood fillers with a thermoplastic polymer gives a light material, having good mechanical strength and suitable for reprocessing [7, 8]. Authors [9] investigating the mechanical properties of a material based on a linear low density polyethylene filled with crumbled pumpkin seed husks found their good suitability as a filler, which may predispose such materials to various applications.

The aim of the study was to determine the effect of powdered fillers obtained from renewable resources, such as wheat bran, pumpkin seed husks and peanut shells on selected mechanical properties of low-density polyethylene (LDPE) filled with these natural fillers. In particular, the study examined the relationships between the basic strength and thermal properties of the produced injection moulded pieces and filler contents.

IOP Conf. Series: Materials Science and Engineering **600** (2019) 012020 doi:10.1088/1757-899X/600/1/012020

2. Experimental

2.1. Material
Test specimens were made of powdered low-density polyethylene (LDPE) Dowlex 2631.10UE manufactured by the DOW Chemical Company (Schkopau, Germany). This polymer is used for producing thin-walled parts by rotational casting and high dimensional accuracy parts by injection molding. Table 1 lists the properties of the tested polymer after the specifications provided by the manufacturer.

The first filler used in the study was wheat bran, i.e. the coat of wheat grains, coming from a local mill. The second natural filler used in the research were pumpkin seed husks obtained from the plant dealing with the purification and sale of pumpkin seeds. The third natural filler used in the research were peanut shells obtained as a result of separating the edible part of the nut.

For a better mixing effect an adhesion promoter in the form of a carbofunctional silane called aminopropyltriethoxysilane was used.

Table 1. Selected properties of Dowlex 2631.10UE.

Property	Value
Density 23°C, kg/m^3	935
Melt flow rate (230 °C; 2.16 kg), g/10 min	7
Tensile stress at yield, MPa	17.8
Tensile strain at yield, %	419
Shore hardness, °Sh D	56
HDT temperature, B (0.45 MPa) °C	52
Vicat softening temperature (A120 (120°C/h 10N), °C	115
Melting temperature, °C	124

2.2. Test stand
The tests were performed using the ARBURG ALLROUNDER 320C 500 – 170 single-screw injection molding machines (Loßburg, Germany), provided with a two-cavity mold for manufacturing standard specimens in compliance with ISO 527-1:2012. The injection molding machine has a cylindrical screw with a diameter of 30 mm and the length to diameter ratio of 20. The highest peripheral screw speed equals 42 m/min and its torque is up to 250 Nm. The composite polymer was injected into the mold at the pressure of 100 MPa and the holding pressure equaled 85 MPa.

2.3. Methods
The process of preparing organic fillers covered their fragmentation into a powder and separation of fractions of a given grain size using sieves with mesh sizes of 0.2mm, 0.8mm and 1.2mm. As a result, three fractions with grain sizes comprised in intervals (<0.4mm, 0.4-0.8mm, 0.8-1.2mm) were obtained. Before starting the injection molding machine, mixtures of low density polyethylene were prepared with organic fillers from renewable sources, i.e. wheat bran, pumpkin seed husks and peanut shells with different mass content of fillers equaling 0, 5, 10, 15 and 20%.

The static tensile tests were performed on produced composite moldings using a standard testing machine, Z010 AllroundLine from Zwick Roell (Ulm, Germany). The Z010 testing machine has the maximum tensile force of up to 10kN and the tensile rate up to 2000mm/min. Static tensile tests were performed using the testing machine on 10 injection moldings in compliance with the ISO 527-1:2012 standard.

Hardness measurements were conducted by Shore method using the ART.13 hardness tester manufactured by Affri System Hardness Testers (Induno Olona, Italy). Hardness measurements were conducted in accordance with the procedure described in the ISO 48-4:2018 standard. Heat deflection temperature (HDT) and Vicat softening point were measured using the CEAST HV3 from INSTRON

CEAST Division. The temperature measurements were made in compliance with the procedures described by two standards: PN-EN ISO 75 (HDT) and PN-EN ISO 306 (Vicat).
The examination of morphology of the specimen cross sections was performed using the Nikon Eclipse LV100ND microscope (Warsaw, Poland), equipped with the DS-U3 camera and NIS-Elements AR 4.20.00 software.

3. Results and discussion

3.1. Strength tests of produced injection moldings

On the basis of the obtained results of strength measurements in the static stretching test, diagrams of relevant relationships were made. An example diagram showing the dependence of the Young's modulus from the content of natural fillers and the size of their grains is shown in Figure 1.
All polymer mixtures filled with organic fillers showed an increase in Young's Modulus, the highest at 20% by weight of the filler. This shows that the stiffness of the polymer has increased, which is probably due to the limited mobility of polymer chains because of the presence of a natural filler [10]. An important role is also played by the size of grains at the phase boundaries, the LDPE polymer matrix - the surface of the filler grains [11]. The larger the size of the filler grains, the higher the value of the Young's modulus.

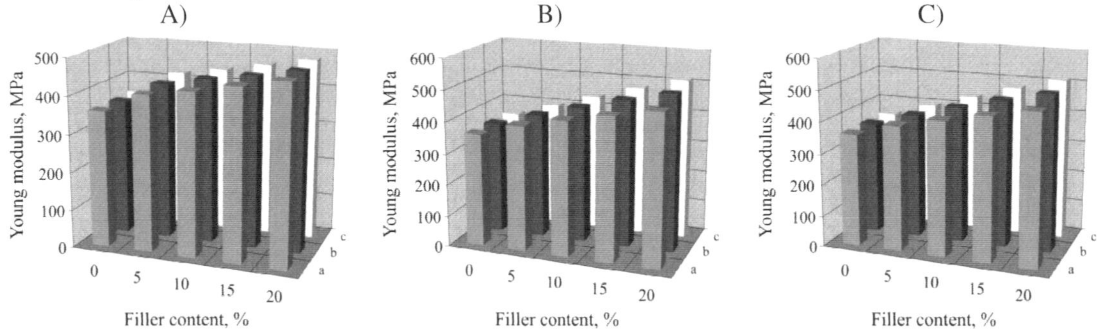

Figure 1. Graph of Young's modulus of low-density polyethylene filled with powdered organic fillers versus various contents and grain sizes: a) grain size from 0 to 0.4 mm, b) from 0.4 to 0.8 mm, c) from 0.8 to 1.2 mm; A) peanut shells, B) pumpkin seed husks, C) wheat bran.

It was also found that the addition of 20% filler causes a decrease in the value of the highest tensile stress (Fig. 2). This is probably due to the matrix being a polyolefin polymer which contributes to the weakening of the adhesion of the hydrophilic organic fillers to the hydrophobic polymer matrix. The structure of most natural fillers favors the increase of water absorption, which results in the variability of mechanical properties [12].

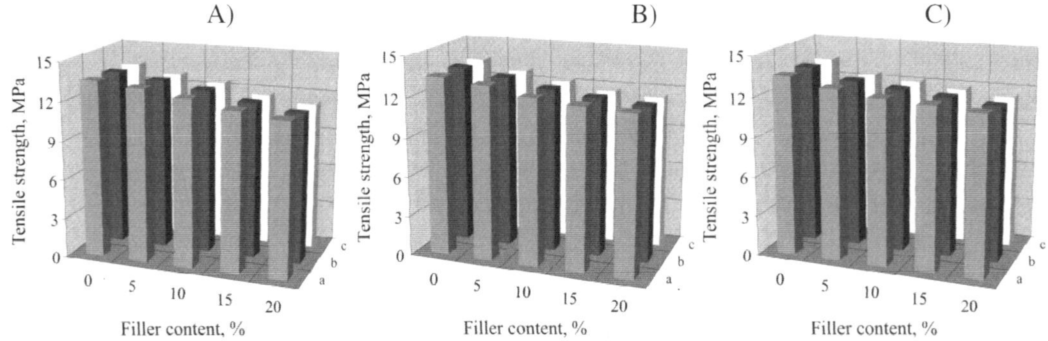

Figure 2. Graph of tensile strength of low-density polyethylene filled with powdered organic fillers versus various contents and grain sizes: markings as in Figure 1.

Analogous tendencies of change of the described property were obtained in other works, in which the effect of various natural fillers on the mechanical properties of polymer compositions such as rice bran, maize starch and potato starch [13] was investigated.

3.2. Hardness of injection moldings

For the purpose of investigations to determine the effect of mass content and various sizes of natural filler grains on the hardness of the obtained injection moldings, Shore D hardness measurements were made. The results of hardness tests are presented in Figure 3 in the form of dependence of hardness expressed in Shore D on mass content and various sizes of filler grains used for testing. When analyzing the graph, it should be noted that the grain sizes used and the mass content of natural fillers significantly affect the hardness of polymer mixtures.

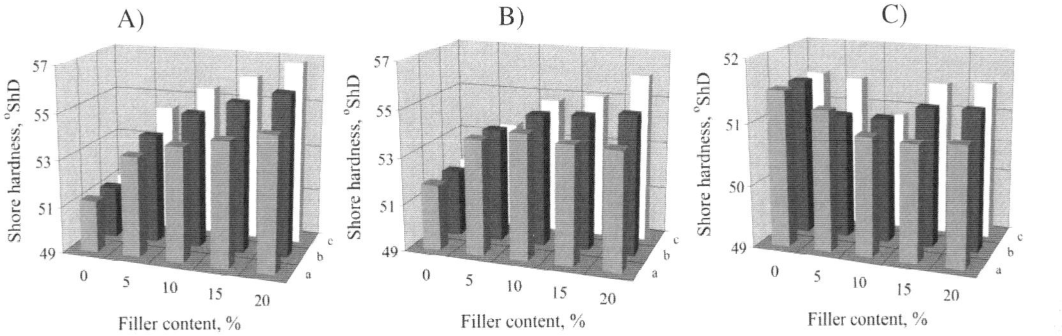

Figure 3. Dependence of hardness of low-density polyethylene filled with powdered natural fillers on their various contents and grain sizes: markings as in Figure 1.

3.3. Vicat softening point and Heat deflection temperature (HDT)

No rule was defined that would describe the effect of the presence of powder fillers on the Vicat softening point (Figure 4A). Temperature changes depend on the type of filler and its interaction with the polymer matrix. In the analyzed case, it is difficult to talk about the influence of the filler on the softening point due to the fact that the percentage values of temperature changes reach a maximum of 1% of the initial value.

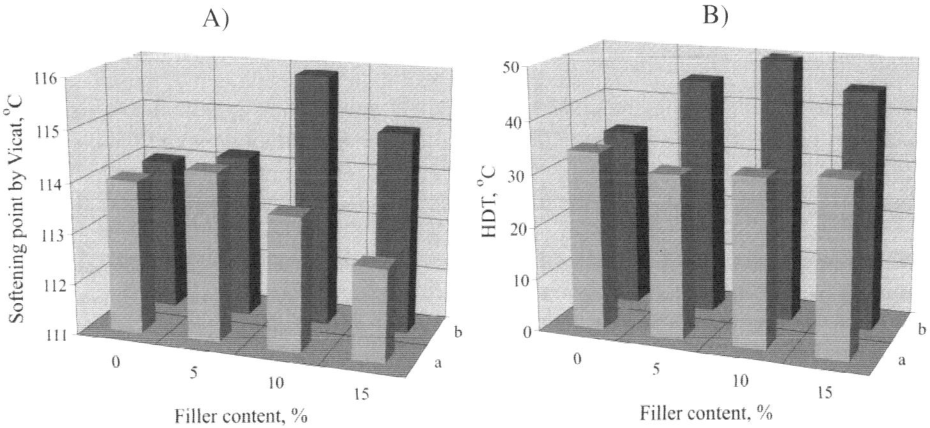

Figure 4. Dependence of: A) Vicat Softening Temperature, and B) Heat Deflection Temperature of LDPE filled with powdered natural fillers in the form of a) wheat bran and b) pumpkin seed husks on various contents.

The graphic dependence of the heat deflection temperature on the content and type of the filler is shown in Figure 4B. The heat deflection temperature of the injection moldings tested shows a decreasing tendency when increasing the content of the natural filler in the form of wheat bran. The addition of powder filler in the form of pumpkin seed husks to low density polyethylene results in an increase in the tested property and then a decrease. The probable cause of the decrease in the heat deflection temperature of the obtained samples with the addition of wheat bran may be the poor adhesion of the filler grain to the polymer matrix.

3.4. Morphology of produced polymer compositions

As a part of the conducted tests, microscopic tests of the cross-section of the obtained polymer mixtures containing 5, 10, 15 and 20% of natural fillers were carried out. Only the pictures of morphology of the injection moldings examined containing pumpkin seed husks were presented in this article (Fig. 5).

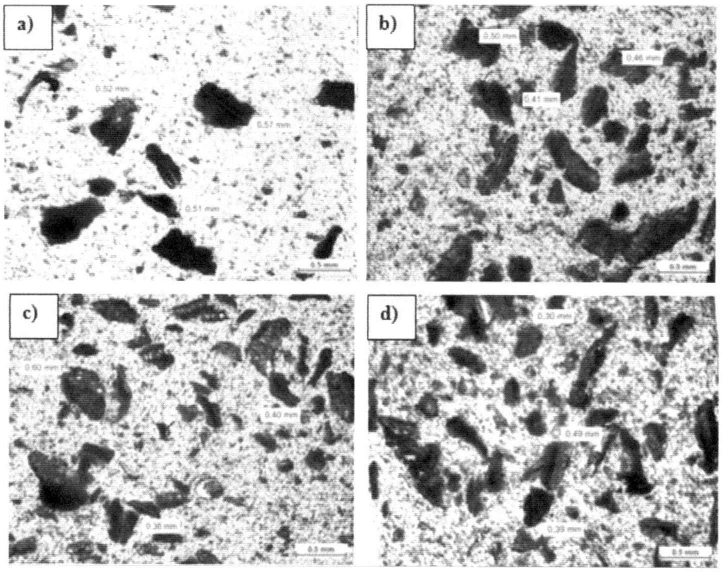

Figure 5. Microscopic photos of the cross-section of tested polymer mixtures filled with powdered pumpkin seed husks of the grain size 0.4-0.8 mm: a) LDPE + 5% natural filler; b) LDPE + 10% natural filler; c) LDPE + 15% natural filler; d) LDPE + 20% natural filler.

The analysis of the morphology showed an uneven distribution of the added natural filler. Particles of varying sizes and agglomerates can be observed. Micrometric pumpkin seed husks particles with probably higher surface energy than surface energy of low density polyethylene tend to agglomerate in such medium (Figures 5b, 5c and 5d), which is an unfavorable phenomenon, because it reduces the contact surface between the filler and the polymeric material, which results in reduction of adhesion.

4. Conclusions

Modification of polymer materials with various types of natural fillers brings with it many changes starting from the course of the processing process, mechanical and thermal properties, to the morphology of the obtained products. Analysis of changes in the properties of such modified polymers is important from the point of view of the applicability of the polymer composition.

On the basis of the research on the properties of a polymer mixture of low density polyethylene with natural fillers, it has been found that they are suitable for use as a filler for this material. In the case of Young's modulus tests, there was a significant increase in this value, and in the case of hardness tests, a slight increase in the tested property along with an increase in the content of fillers and grain size. VST and HDT do not show significant changes in the range of applied grain sizes and the tested mass filler content.

Acknowledgments
The project/research was financed in the framework of the project Lublin University of Technology-Regional Excellence Initiative, funded by the Polish Ministry of Science and Higher Education (contract no. 030/RID/2018/19)".

References

[1] Sasimowski E., Sikora J., Królikowski B.: Effectiveness of polyethylene extrusion in a single-screw grooved feed extruder. Polimery 2014, 59, 6, 505-510.
[2] Alonso, M., Velasco, J. ., de Saja, J.A.: Constrained crystallization and activity of filler in surface modified talc polypropylene composites. European Polymer Journal 1997, 33, 3, 255-262.
[3] Głogowska K., Sikora J.W., Duleba B.: Effect of mechanical properties of metal powder filled hybrid moulded products. Journal of Polymer Engineering 2016, 36, 7, 705-712.
[4] Chmielnicki B., Konieczny J.: Właściwości kompozytów WPC o osnowie polietylenowej napełnionych maczka z łupin orzechów. Przetwórstwo tworzyw 2014, 20, 1, 12-20.
[5] Żenkiewicz M., Richert J., Permeability of polylactide nanocomposite films for water vapour, oxygen, and carbon dioxide, Polymer Testing, 27, 2008, 835–840.
[6] Żenkiewicz M., Richert J., Influence of polymer samples preparation procedure on their mechanical properties, Journal of Achievements in Materials and Manufacturing Engineering, 26, 2, 2008, 155–158.
[7] Cyga R., Czaja K.: Kompozyty polimerowe z napełniaczem roślinnym. Przemysł chemiczny 2008, 87, 9, 932-936.
[8] Głowacka L., Zajchowski S.: Właściwości i przetwarzanie kompozytów polimerowo-drzewnych. Inżynieria i Aparatura Chemiczna 2005, 44, 3s, 26-27.
[9] Głogowska K., Sikora J.W: Properties of moldings prepared from LLDPE pumpkin seed hulls blend. Advances in Science and Technology Research Journal 2017, 11, 4, 318-325.
[10] Rothon R.: Mineral Fillers in Thermoplastics: Filler Manufacture and Characterisation. Advances in Polymer Science 1999, 139, 67-107.
[11] Khan M., Hyder S.A., Haitao Q.: On accelerated flows of a viscoelastic fluid with the fractional Burgers' model. Nonlinear Analysis: Real World Applications 2009, 10, 2286–2296.
[12] Pinheiro L.A., Chinelatto M.A., Canevarolo S.V.: The role of chain scission and chain branching in high density polyethylene during thermo-mechanical degradation. Polymer Degradation and Stability 2004, 86, 3, 445-45.
[13] George J., Kumar R. and others: Rice bran-filled biodegradable low-density polyethylene films: Development and characterization for packaging applications. Journal of Applied Polymer Science 2006, 5, 4514-4522.

IOP Conf. Series: Materials Science and Engineering **600** (2019) 012021 doi:10.1088/1757-899X/600/1/012021

Experimental evaluation of high-strength concrete-filled steel rectangular tube columns under axial compression

Tharangkoon Konkam[1], Jaksada Thumrongvut[1*], Cherdsak Suksiripattanapong[1]

[1]Department of Civil Engineering, Faculty of Engineering and Architecture, Rajamangala University of Technology Isan, Nakhon Ratchasima, 30000, Thailand

*E-mail: jaksada.th@rmuti.ac.th

Abstract. This paper presents the effect of the high-strength concrete on the behavior and strength of concrete-filled steel tube (CFST) columns. To evaluate the structural performance, load-axial shortening responses, failure modes and axial load capacity are performed for ten high-strength concrete-filled steel tube (HSCFT) columns. All of the columns are 750 mm in height, and have compressive strength of 66 MPa. Test parameters included the wall thicknesses of the rectangular tube (3.0 mm, 4.5 mm and 6.0 mm) and three different of rectangular cross-sections (150x50 mm, 150x75 mm and 150x100 mm). From the experimental results, it is indicate that the load-axial shortening response of the columns have a linear behavior up to approximately 80-90% of their axial load capacity. Then, the behavior of the HSCFT columns is nonlinear with a very large deformation before failure. It is also shown that the axial load capacity and ductility of the HSCFT columns are increased significantly compared to the RC columns (reference columns), depending mainly on the wall thickness and cross-section of the hollow steel tube. Finally, by comparing the axial load capacity of the test results with those obtained from the AISC/LRFD design equation, the comparison results show that calculation formula in AISC/LRFD specification can be applied to compute the axial load capacity of HSCFT columns under axial compression.

1. Introduction

Concrete-filled steel tube (CFST) column is a structural member that uses a combination of concrete and cold-formed steel circular and rectangular tubes to provide adequate axial load capacity [1-2]. Concrete cores have the advantages of high compressive strength and stiffness, while thin-wall hollow tubes have the advantages of high strength and better ductility performance [3]. The steel tube serves as reinforcement and permanent formwork for the concrete fill. It is well known that the concrete core delays the inward buckling and temperature rise of steel tube, and significantly enhances resistance and inelastic deformation capacity in compression and combined loads. Also, CFST columns combine steel and concrete in one member resulting in a member that has the beneficial qualities of both materials, such as high ductility and large energy absorption for earthquake resistant properties, economical, reducing labor requirements and permitting rapid construction, when compared to traditional construction methods [4-8].

Although the use of high-strength concrete (HSC) in the construction sector provides significant economic benefits, the understanding of the behavior of high-strength concrete-filled steel tube (HSCFT) columns are still insufficient in comparison with normal strength CFST columns [9]. Consequently, there is a need to extensively investigate the structural performance of HSCFT columns

IOP Conf. Series: Materials Science and Engineering **600** (2019) 012021 doi:10.1088/1757-899X/600/1/012021

under axial compression so that guidelines and recommendations can be developed for design. The purpose of this paper is to present the experimental results on HSCFT columns under concentric axial loading and to compare the obtained axial load capacity with the reference RC columns and the values predicted by AISC/LRFD specification [10].

2. Experimental procedures

2.1. Specimen details

An experimental study was made on 16 rectangular columns, including 10 HSCFT columns and 6 reference high-strength RC columns. All rectangular columns were 750 mm in height. The main parameters were the wall thicknesses of the rectangular tube (3.0 mm, 4.5 mm and 6.0 mm) and three different of rectangular cross-sections (150x50 mm, 150x75 mm and 150x100 mm). Two specimens of each set were tested to determine the average axial load capacity. A summary of the nominal dimensions of HSCFT columns are presented in Table 1. They were classified into 3 groups with H/B ratio. The column numbers were designated as RC(RS)- f_c' - t - H/B, where "RC" or "RS" represents the reference high-strength RC columns and the HSCFT columns, respectively. For example, the column number RS-66-3.0-2.0 was the HSCFT columns, having f_c' = 66 MPa, t = 3.0 mm and H/B ratio of 2.0. Also, the average mechanical properties of the high-strength concrete at the age of 28 days and steel tube were tested according to ASTM procedures, and given in Table 1.

Table1. Dimensions and mechanical properties of column specimens.

Specimen	t (mm)	H (mm)	B (mm)	H/B	H/t	Concrete		Steel Tube		Number
						A_c (mm^2)	f_c' (MPa)	A_{st} (mm^2)	f_y (MPa)	
RC-66-0-1.5	-	150	100	1.5	-	14,548	66.2	-	-	2
RS-66-3.0-1.5	3.0	150	100	1.5	50	13,467	66.2	1,533	382.2	2
RC-66-0-2.0	-	150	75	2.0	-	10,798	66.2	-	-	2
RS-66-3.0-2.0	3.0	150	75	2.0	50	9,959	66.2	1,291	384.5	2
RS-66-4.5-2.0	4.5	150	75	2.0	33.3	9,358	66.2	1,892	388.2	2
RS-66-6.0-2.0	6.0	150	75	2.0	25	8,787	66.2	2,463	395.4	2
RC-66-0-3.0	-	150	50	3.0	-	7,048	66.2	-	-	2
RS-66-3.0-3.0	3.0	150	50	3.0	50	6,209	66.2	1,141	380.8	2

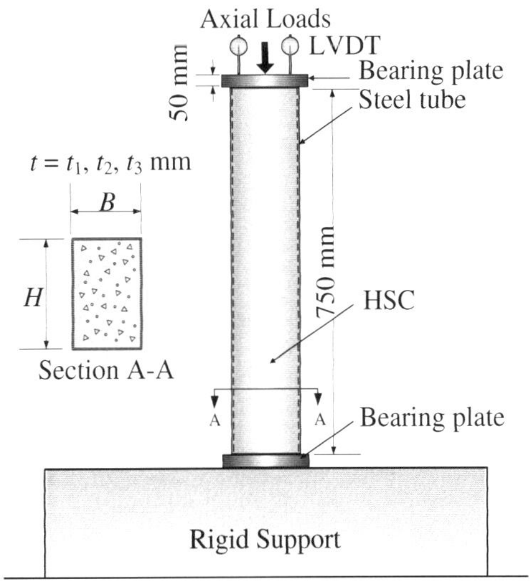

Figure 1. Typical HSCFT column details and test setup.

IOP Conf. Series: Materials Science and Engineering **600** (2019) 012021 doi:10.1088/1757-899X/600/1/012021

2.2. Test setup and instrumentations

Depending on the axial load capacities of each column, axial compressive test of the column specimens were performed by a 2,000 kN TINIUS OLSEN Universal Testing Machine. The axial compressive load was applied to entire section through the bearing plates (applied load to the concrete and steel tube section). The load arrangement and test setup can be seen in Figure 1. Two linear variable differential transducers (LVDT) were used to monitor overall axial shortening. Before the beginning of each test, a preload of approximately 20% of the predicted axial load capacity of the columns was applied in order to reduce the friction between the steel bearing plates and the columns, and to balance the uneven top and bottom surfaces. The specimens were loaded at a very slow rate, was increased at a constant rate up to axial load capacity, such that the axial load, the axial shortening of the columns were automatically recorded by a KYOWA EDX-10 Series data acquisition unit. Additionally, the local tube wall buckling and the axial load capacity were carefully observed.

3. Results and discussion

3.1. Behavior and failure mode

In this paper, the term "short column" refers to a compression member which can attain its axial load capacity, known as the squash loads (P_{Test}), without buckling of the columns. The typical structural behavior of the tested reference high-strength RC column and HSCFT column are represented in Figure 2 by relations between the axial load and axial shortening. From Figure 2, it can be seen that the HSCFT column showed a linear elastic behavior up to 80-90% of their axial load capacity. In elastic stage, the deformation of HSCFT column was too small and invisible. After that, as the load continued to increase and reached ultimate load, the curves are gradually becoming nonlinear due to the start of the cracking of the concrete underneath the applied load and the local yielding of the steel tube. Then, it is seen that nonlinear behavior of the columns is the strain-softening type, in which the load began to decrease and the axial shortening increased rapidly. Finally, the outward local wall buckling of the steel tube in the areas near the ends of the columns are investigated and the axial load capacity is recorded. Figure 3 shows a typical failure mode of the HSCFT column. The failure mechanism of the columns was identified as the crushing of concrete core and the yielding of steel tube. This indicates that the concrete core in these locations was contained by the steel tube, in turn providing large axial deformability to the columns. Hence, the HSCFT column represented well post-yield behaviour.

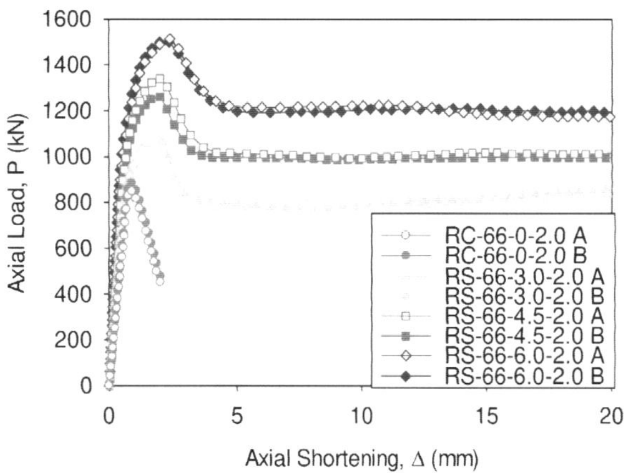

Figure 2. Axial load versus axial shortening

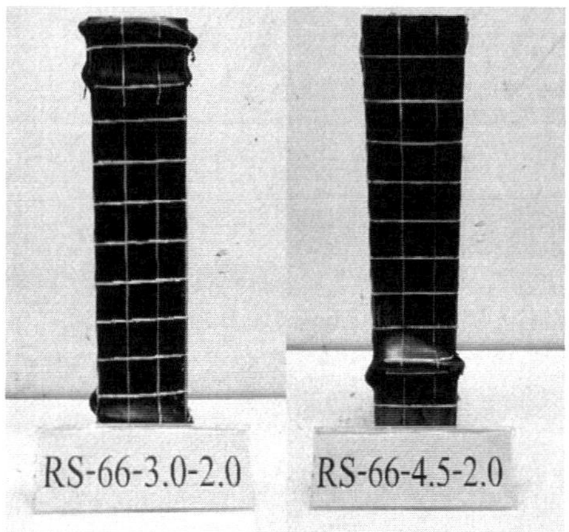

Figure 3. Failure modes of HSCFT column

3.2. Axial load capacity and comparison with AISC/LRFD

Figure 4 presents the comparisons of the obtained axial load capacity (P_{Test}) of the HSCFT column. It can be seen from the ratio of the axial load capacity of the HSCFT column to the axial load capacity of the reference high-strength RC column ($P_{Test,ref}$) or P_{Test} / $P_{Test,ref}$ ratio that the HSCFT column with t = 3.0 mm, 4.5 mm and 6.0 mm have larger P_{Test} than the $P_{Test,ref}$ of the reference CLC column by 37%, 64% and 89%, respectively. This indicates that the axial load capacity increased with the increase of thickness [11].

In addition, AISC Load and Resistance Factor Design (LRFD) is the most completed specification in composite members [3], and covers composite columns, including the CFST column. From the analytical results, the comparison of axial load capacity between the test values and the LRFD design equation are presented in Figure 5. The comparison of results shows that LRFD closely predicts with a difference of 5-21%. This indicates that the LRFD design equation can adequately predict the axial load capacity of the HSCFT column.

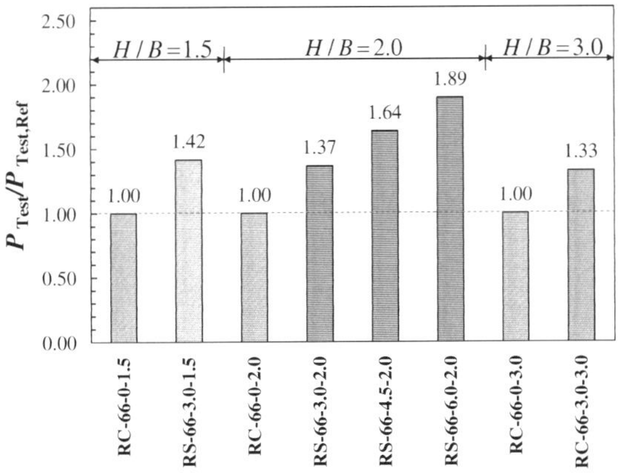

Figure 4. Ratio of P_{test} / $P_{test,ref}$ of test columns

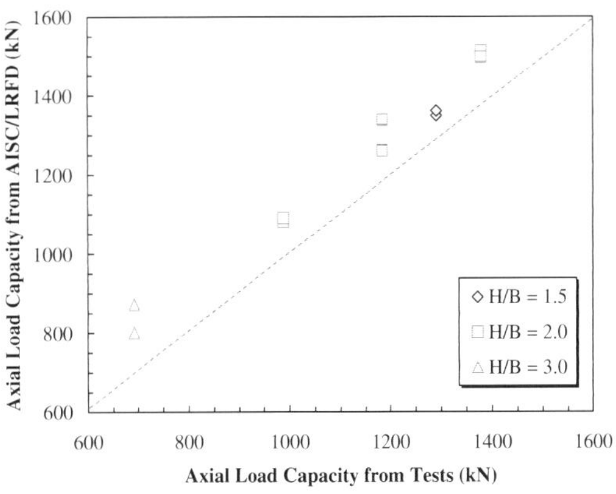

Figure 5. Comparison of the axial load capacity obtained from experiments and AISC/LRFD Eq.

4. Conclusions

The conclusions from the experimental studies are summarized as follows.

1) The structural performance of HSCFT column have a linear elastic behavior up to 80-90% of their axial load capacity. Then, the behavior of the columns is gradually becoming nonlinear with the strain-softening type. Failure mechanism of the columns is identified as the crushing of concrete core and the localized buckling of steel tube.
2) The typical failure mode of the HSCFT column is progressive mode with a very high axial deformability.
3) The encasing effect is more effective for the HSCFT column with thicker wall thickness than those with thinner wall thickness.
4) Good agreement is evaluated between the predicted axial load capacity using the AISC/LRFD design equation and the experimental results.

Acknowledgments

The authors would like to acknowledge the financial support provided by the Thailand Research Fund (TRF): Research and Researchers for Industries (RRI) which is a part of the research project "Development of High-Strength Concrete-Filled Steel Rectangular Tubular Columns for Applications in Building and Infrastructure". The authors are grateful to Faculty of Engineering and Architecture, Rajamangala University of Technology Isan for facilities and equipment provided.

References

[1] Han, L.H., Li, W. and Bjorhovde, R. (2014). Developments and advanced applications of concrete-filled steel tubular (CFST) structures: Members. *Journal of Constructional Steel Research*, 100: 211-228.

[2] Thumrongvut, J. and Tiwjantuk, P. (2018). Strength and axial behavior of cellular lightweight concrete-filled steel rectangular tube columns under axial compression. *Materials Science Forum*, 941: 2417-2422.

[3] Saw, H.S. and Liew, J.Y.R. (2000). Assessment of current methods for the design of composite columns in buildings. *Journal of Constructional Steel Research*, 53(2): 121-147.

[4] Seangatith, S. and Thumrongvut, J. (2011). Behaviors of square thin-walled steel tubed RC columns under direct axial compression on RC core. *Procedia Engineering*, 14: 513-520.

[5] Ananthi, B.G. and Knight, S. (2014). Experimental and theoretical study on cold-formed steel box stub columns under uniaxial compression. *Arabian Journal for Science and Engineering*, 39(10): 6983-6993.

[6] Lu, Y., Li, N., Li, S. and Liang, H. (2015). Behavior of steel fiber reinforced concrete-filled steel tube columns under axial compression. *Construction and Building Materials*, 95: 74-85.

[7] Ding, F., Lu, D., Bai, Y., Zhou, Q., Ni, M., Yu, Z. and Jiang, G. (2016). Comparative study of square stirrup-confined concrete-filled steel tubular stub columns under axial loading. *Thin-Walled Structures*, 98(B): 443-453.

[8] Thumrongvut, J., Seangatith, S., Siriparinyanan, T. and Wangrakklang, S. (2016). An experimental behaviour of cellular lightweight concrete-filled steel square tube columns under axial compression. *Materials Science Forum*, 860: 121-124.

[9] Liu, D., Gho, W.M. and Yuan, J. (2003). Ultimate capacity of high-strength rectangular concrete-filled steel hollow section stub columns. *Journal of Constructional Steel Research*, 59(12): 1499-1515.

[10] ANSI/AISC 360-10. (2010). *Specification for structural steel buildings*. American Institute of Steel Construction, Chicago, IL.

[11] Thumrongvut, J. and Seangatith, S. (2016). Axial load capacity of cellular lightweight concrete-filled steel square tube columns. *Central Europe Towards Sustainable Building (CESB 2016)*, Czech Republic, 22-24 June 2016: 1312-1319.

Carbide tool life prediction and modeling in SiCp/Al turning process via artificial neural network approach

Rashid Ali Laghari[1]*, Jianguang Li[1], Asif Ali Laghari[2], Mozammel Mia[3], Shu-qi Wang[1], Wang Aibo[1], Poonam K. K.[4]

[1]School of Mechatronics Engineering, Harbin Institute of Technology, Harbin, 150001, China
[2]School of Computer Science & Technology, Harbin Institute of Technology, Harbin, 150001, China
[3]Mechanical and Production Engineering, Ahsanullah University of Science and Technology, Dhaka 1208, Bangladesh
[4]Department of Science and Technology, Indus University, Karachi, Pakistan

[1]rashidali@stu.hit.edu.cn; [1]mejgli@hit.edu.cn; [2]asiflaghari@hit.edu.cn; [3]mozammelmiape@gmail.com; [4]kkpoonam29@yahoo.com; [1]15b308001@hit.edu.cn; [1]wangaibo1991@163.com
Corresponding author email: [1]rashidali@stu.hit.edu.cn, rashidalilaghari@gmail.com

Abstract. The experimental research and theoretical analysis of tool life and attached wear with the tool during turning operation of SiCp/Al 45%vol: fraction has been carried out. This research proposed the machining factors affecting the wear of carbide tool and the laws of wear affecting the life expectancy of the tool. Artificial neural network (ANN) was utilized for analyzing the cutting process, the training of ANN achieved by back-propagation on the basis of three input parameters such as cutting speed, feed and depth of cut. The wear and damage of cutting tools in the machining process of composite materials also depend on SiC reinforced particles size and volume. The effects of workpiece material components and different cutting process parameters on the tool wear mechanism were thoroughly analyzed and measured. The main wear form on the cutting tool and its major causes for different wear patterns were recognized as adhesive and abrasive, for such phenomena the cutting speed held as a most influencing factor.

Keywords: SiCp/Al Metal Matrix Composite; tool life; wear mechanism; turning process; Artificial Neural Network; Modeling.

1. Introduction

Metal matrix composites (MMCs) containing reinforced components, it is a fresh branch of composite materials. MMC materials have high specific strength and modulus along with the excellent physical and mechanical properties. MMC materials have high performance; contain low weight in comparison to volume with other metals [1]. Major applications of MMCs has found in the development of aerospace and automotive technology due to characteristics of the complex structure, and high precision. MMCs are also known as aerospace material due to its comprehensive performance in the development of spacecraft structural parts. Metal matrix composite materials have achieved much more respect in both military and civilian industrial applications [2].

In spite of being composite materials exhibiting the wide range of application, however, their machining cost has not been affordable for general required dimensional accuracy of the component. Several researchers found that the MMCs machinability involves in sudden tool failure and high tool flank wear with an adverse effect on component surface quality due to the presence of SiC hard particles. It has revealed from the literature that several cutting tools fail to machine MMCs [3].

The material of the employed cutting tool should be more in hardness than metal matrix composite materials for the machining process. The cutting tool should be strong enough to endure the wear mechanism during the operation for a certain period to give suitable life to the cutting tools. From different studies, it has seen that the cubic boron nitride (CBN), tungsten carbide (WC), diamond coated carbide tools and polycrystalline diamond cutting tool (PCD), are decent in machining of composite material in accordance to their respective characteristics. It is also found out that these tools have performed well in a quality surface generation at moderate cutting conditions [3, 4, 5].

Yanming and Zehua study the metal matrix composite materials affecting features on the tool wear. From machinability outcomes, where it reveals that the harsh nature of MMC reinforced particle's size and volume fraction in a matrix are the most impelling aspects affecting tool life [5]. Similarly, Kilickap et al. have presented the results of titanium carbide (K10) tool wear in the cutting process of SiCp/Al MMC. The clarifications show the growth in tool wear during increasing the cutting speed and due to an adverse effect of hard SiC particles on the insert. However, the surface roughness effected by both increasing cutting speed and feed rate [6].

This attempted work conducted to understand the carbide tool wear mechanisms and behavior in CNC lathe turning of Al/SiCp45% vol: metal matrix composite. The wear model of carbide tool has established on the basis of artificial neural network data optimizing technique a general equation involving machining conditions cutting speed, feed rate and depth of cut.

2. Artificial Neural Network (ANN)

Artificial Neural Network (ANN) is an evolving information processing technique, which is studied by several scientists and employed in various research and technical fields. Features of ANN somehow resemble with the human brain, it mimics the information processing, storage and it's similar to the retrieval functions of the human brain [7]. ANN possesses intelligent processing functions of memorizing, learning and calculating, due to its ability of self-organization and adaptation. After several years of research and development, there are many numbers of neural network models in ANN, which roughly categories into three major sets such as self-organizing neural network, Feed-back neural network and feed-forward neural network. In the area of the metal cutting process, the ANN is employed to measure and understand the machine tool behavior and analysis the error. The core application of ANN in the metal cutting process is to evaluate the effect of machining parameters on tool wear, optimization of parameters and reduction of errors in machining operation [8]. In this experimental work, research is conducted on the lathe turning to measure and analyze the tool wear and its mechanisms in the cutting process of SiCp/Al reinforced metal matrix composite material.

In this specific research the architecture of 3-n-1, a feed forward back-propagation multi-layer neural network is employed for analysis of tool wear. The ANN model is developed in the ANN tool-box in MATLAB R2015a. Developing the productive ANN model with less number of data points are most challenging task. However, developing model with less data point determine the capability of this particular technique in formulating the ANN tool wear model. From the available 16 experimental data sets, randomly nominated 12 sets are taken to build the model and remaining of the 4 data sets are used for the validation of the developed model. Fig: 1 shows the ANN configuration used for training in this research.

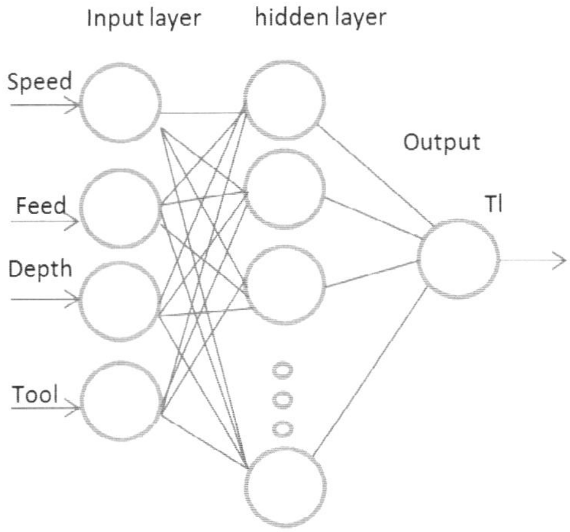

Figure 1. ANN configuration used for training

3. Experimental conditions

In this experiment, the orthogonal machining progression is accomplished on composite materials with three different cutting responses namely cutting speed, feed rate and depth of cutting. The experimental sample containing the different size such as measured length of 98.9 mm and measured diameter possess 48.2 mm. wear progression on cutting tool was measured after completing every single cutting operation for all number of cutting operation. There are 16 sets of cutting operations were performed in dry cutting conditions with the non-coated carbide tools. The average height 0.3 mm is counted as an of the flank wear to be considered as a worn edge with respect to ISO 3685 (ISO, 1993).
The machining parameters utilized in this operation for turning workpiece are detailed in Table 1. However, the experimental setup for turning operation on SiCp/Al is illustrated in fig: 2.

Table1. Cutting Parameters

Cutting speed (m/min)	6.283	12.566	18.85	25.133
Feed rate (mm/r)	0.01	0.015	0.02	0.025
Depth of cut (mm)	0.2	0.5	1	1.5

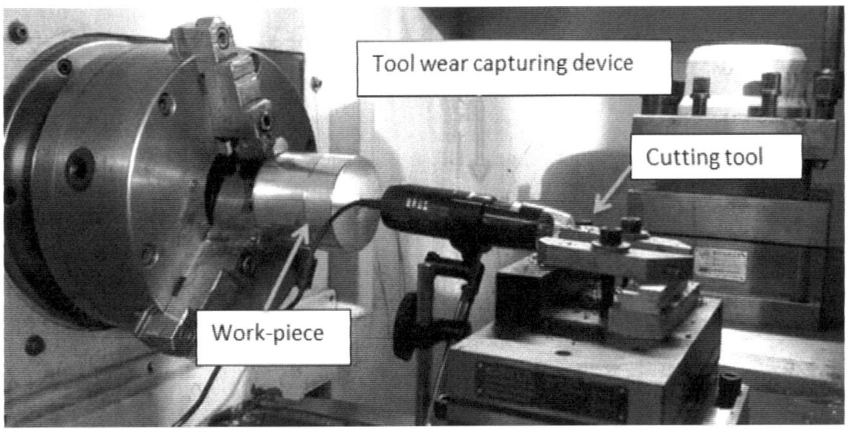

Figure 2. Experimental setup of turning of SiCp/Al

IOP Conf. Series: Materials Science and Engineering **600** (2019) 012022 doi:10.1088/1757-899X/600/1/012022

4. Estimation of carbide tool wear in turning process of SiCp/Al

4.1 Effect of SiCp/Al workpiece material properties on tool wear

It has been clarified in several studies that the more appropriate tool material for processing SiCp/Al metal matrix composites is a PCD cutting tool. PCD cutters wear mechanisms and amount are much less than cemented carbide tools, however, carbide tools have good economic machining performance, which can be used as tool materials at lower cutting parameters for MMCs, and are also very suitable for rough machining of particle reinforced aluminum matrix composites [4].

When machining SiCp/Al used to performed, the tool wear and wear mechanism are related to the volume and size of the reinforcing particles in the material. The greater the particle content and size in the composite, the more severe the tool wear. It can be seen from the experimental results that in the processing of particle reinforced aluminum matrix composites, the tool wear phenomenon varies with different component materials due to the difference in internal particle size and particle content [9].

The following fig: 3 shows the different patterns of carbide tool wear, and on the other hand fig: 4 show the comparisons between the experimental values and ANN calculated values of the tool life. The fig: 5 shows the tool life error percentange of ANN caluculated values with respective to obtained experimental data.

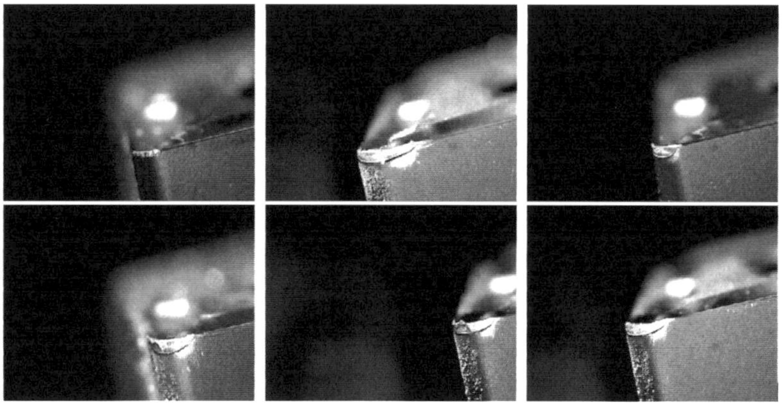

Figure 3. Different tool wear patterns

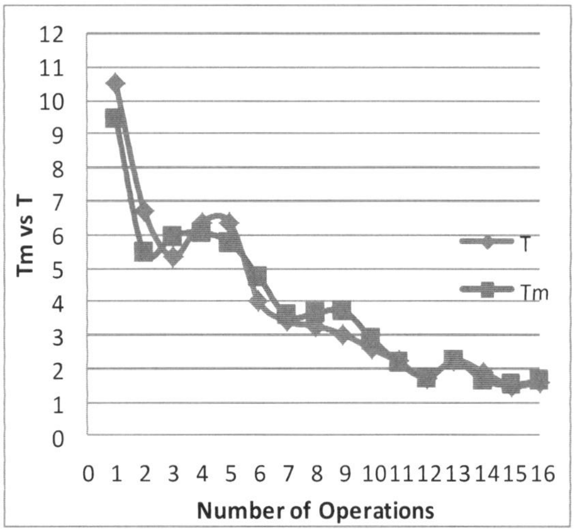

Figure 4. Experimental Tool life (T) vs ANN Tool life(Tm)

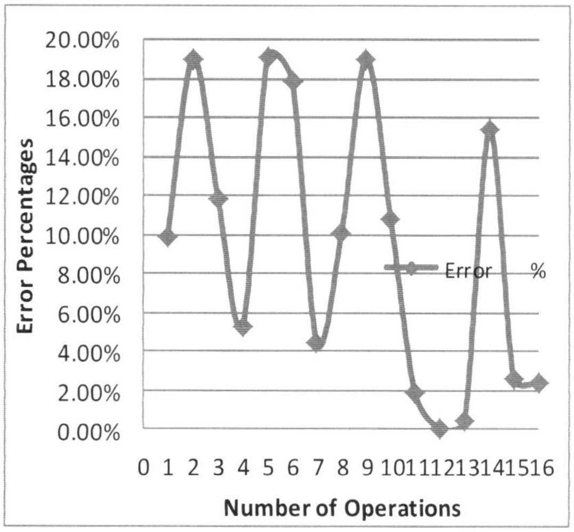

Figure 5. ANN measured Error %

5. Conclusions

This experimental research thoroughly analysis the tool wear law, mechanism and morphology in the cutting process of SiCp/Al reinforced metal matrix composites.

When using carbide cutting tool in turning operation of SiCp/Al reinforced metal matrix composites, the tool wear shows the increasing phenomena with the higher cutting speed.

Tool wear morphology reveals that the flank wears of the carbide tool is severe and most common during the turning process of composite material. The different damage mechanisms were studied on the flank of the tool; the core causes which affect the flank wear is the adhesive and abrasive wear patterns. Reason for the tool flank wear is due to the low strength of tool surface after doing some machining that produces microcracks, and SiCp/Al materials particles bonded to the rack face of the tool and therefore damaged the carbide tool.

The calculated results of ANN model of tool life are very close to the experimental results. The ANN model successfully predicted the experimental results with minimum error.

For future research work employing the sensors for gathering tool wear data should be the effective as compare to and analyze the obtained data. There is a necessity of effective tools, techniques and methods for gather data during operation. It can provide more precise and accurate results as well as fruitful for analyzing data.

References

[1] Nicholls, C.J., Boswell, B., Davies, I.J. and Islam, M.N., 2017. Review of machining metal matrix composites. The International Journal of Advanced Manufacturing Technology, 90(9-12), pp.2429-2441

[2] Li, J. and Laghari, R.A., 2019. A review on machining and optimization of particle-reinforced metal matrix composites. The International Journal of Advanced Manufacturing Technology, 100.9-12: pp.2929-2943.

[3] Davim JP., 2011.Machining of metalmatrix composites. Springer, London

[4] Laghari, R.A., Li, J., Xie, Z. and Wang, S.Q., 2018. Modeling and Optimization of Tool Wear and Surface Roughness in Turning of Al/SiCp Using Response Surface Methodology. 3D Research, 9(4), p.46. doi10.1007/s13319-018-0199-2

[5] Yanming, Quan, and Zhou Zehua., 2000. "Tool wear and its mechanism for cutting SiC particle-reinforced aluminium matrix composites." Journal of materials processing technology 100, no. 1:pp. 194-199.

[6] Kılıckap, E., O. Cakır, M. Aksoy, and A. Inan., 2005. "Study of tool wear and surface roughness in machining of homogenised SiC-p reinforced aluminium metal matrix composite." Journal of Materials Processing Technology 164:pp. 862-867.

[7] Karabulut, Ş., 2015. Optimization of surface roughness and cutting force during AA7039/Al2O3 metal matrix composites milling using neural networks and Taguchi method. Measurement, 66, pp.139-149.

[8] Mia, M., Khan, M.A. and Dhar, N.R., 2017. Study of surface roughness and cutting forces using ANN, RSM, and ANOVA in turning of Ti-6Al-4V under cryogenic jets applied at flank and rake faces of coated WC tool. The International Journal of Advanced Manufacturing Technology, 93(1-4), pp.975-991

[9] Laghari, R.A., Li, J., Laghari, A.A. and Wang, S.Q., 2019. A Review on Application of Soft Computing Techniques in Machining of Particle Reinforcement Metal Matrix Composites. Archives of Computational Methods in Engineering, pp.1-15.

IOP Conf. Series: Materials Science and Engineering **600** (2019) 012023 doi:10.1088/1757-899X/600/1/012023

The development measuring skill for the thermal conductivity of heat pipe, graphite sheet and vapour chamber

Wei-Keng Lin[1], Wen-Hua Zhang[1], Ching-Huang Tsai[1], Shao-Wen Chen[1]

[1]Engineering & System Science Dept., National Tsing-Hua University, No.101, Sec.2, Kuang Fu Rd., Hsinchu, Taiwan, 30013

wklin@ess.nthu.edu.tw

Abstract. This research designed a set of experimental instruments (TDMI) to measure the thermal diffusivity α of the material. TDMI is calibrated by pure substance such as copper, tin, lead, Aluminium 6061 etc., after done the calibration, the thermal conductivity of heat pipe 、 graphite sheet and vapour chamber also be measured. The repeatability error in this experiment, are within 5%. Compared with the thermal diffusivity data of those the pure substance from handbook while for the accurate relative error were within 10% as well.

1. Introduction

D5470 is useful for thermal conductivity measurement of pure substances such as copper, aluminum or homogeneous substances such as TIM. This K value is fix regardless of the X, Y and Z directions. [1, 2]. However, the K value of a non-uniform two-phase flow heat-conducting material such as a heat pipe, a vapor chamber or a brittle material such as graphite sheet cannot be obtained by a simple Fourier law. The heat transfer principle of graphene and vapor chamber is the same as that of heat pipe but the direction of heat conduction is different. The heat transfer mode of heat pipe is one-dimensional, while the heat transfer mode of vapor chamber could be two-dimensional mode. The thermal conductivity k represents whether the material is a good heat conductor, and the large k value represents the larger the power value that the object can transmit, but the time required to reach this final temperature cannot be determined. The thermal duffusivity α (cm^2/s) of the material represents the area where the heat can diffuse in the same power and unit time. Thermal duffusivity α represents the time required to reach the final temperature of the substance is shorter. The k value is the most direct description of the thermal conductor behavior of the material. Most of it measured by Fourier's theorem. However, Fourier's law is achievable for applied only solid heat conduction with the heating area must be the same as the heat dissipation area. For non-uniform device such as heat pipes or vapor chamber, the function of heat dissipated utilize the latent heat of working fluid, the convection heat transfer. In addition, a fragile material such as graphite sheet, or the heating area and the heat dissipating area are different; there exist a thermal diffusion problem; for measuring the thermal conductivity, k of these particular material, Angstrom method will be a good method. Angstrom's theory is input a heat to a one-dimensional strip test object in a periodic pattern. When the periodic sinusoidal is propagated from one side to another side, by measuring the distance and the wave amplitude of these two measuring points, the thermal diffusivity α[3-6] of the object to be tested can be inferred. Angstrom is one-dimensional control volume governing equation can be expressed as equation (1). After the calculation of the separation variable is simplified, the thermal diffusivity of the object to be tested be able calculated by equation (2). According to the Angstrom method, the thermal diffusivity α_{1D} of the one-dimensional test object can be calculated by using the temperature change M, N, the delay time Δt and the distance L between T_c and T_x in the heat balance process. Table 1 shows the symbols of the Angstrom method. In the heat transfer analysis, the thermal diffusivity α_{1D} is the ratio of the thermal conductivity K to the density ρ and the specific heat C, as shown in equation (3).

$$-\left(\frac{\partial T}{\partial t}\right)+\alpha_{1D}\left(\frac{\partial^2 T}{\partial x^2}\right)-\frac{h}{B\rho C_P}(T-T_a)=0 \tag{1}$$

$$\alpha_{1D}=\frac{L^2}{2\Delta t(\ln\frac{M}{N})} \tag{2}$$

$$\alpha_{1D}=\frac{k_{eff}}{\rho C} \tag{3}$$

Table1. Nomenclature of Angstrom method

symbol	description	symbol	description
2B	Thickness of test sample (cm)	α_{1D}	Thermal diffusivity of one dimension (cm^2/s)
A_S	Surface of test sample (cm^2)	C	Specific heat of test sample (J/g-k)
L	Distance between T_c and T_x (cm)	Δt	The delay time of sine wave cycle from T_c to T_x (s)。
M/N	Amplitude ratio of T_c to T_x	k_{eff}	The effective thermal conductivity (W/cm-k)
ρ	Density of test sample (g/cm^3)	M	Amplitude of the sine wave temperature change of T_c
		N	Amplitude of the sine wave temperature change of T_x

2. Experimental device and system

2.1. Thermal diffusivity (α) measuring instrument TDMI

Figure 1 is schematic diagram of thermal diffusivity measurement [7], the heat source input sine wave heating power, and the control method is to use the thermoelectric chip to achieve a periodic heating sine wave temperature curve. The test piece is placed on the heated copper block and two temperature-detecting points T_c and T_x are respectively installed on the piece to be tested. It should be notice that the heating method must be a point heat source; the contact point of the test sample and the copper block should be as small as possible. The thermal diffusion measurement platform TDMI is mainly composed of two parts: (1) the power supply and (2) the thermal diffusion-measuring platform of the instrument shown in Figure 2.

2.2. Calibration of TDMI

In order to verify the accuracy of the TDMI, standard calibration experiments must carried out. The calibration based on pure metals such as copper, tin, lead and aluminium. The standard values of the thermal diffusivity of these metals be able to find in the literature. After measuring the thermal diffusivity α of these metals by the instrument, and comparing with the standard value, the reliability and accuracy of TDMI should verify. The data measurement is in the form of full blind test. First, it must be determined that the experimental data is repeatable. Therefore, it is necessary to perform repeatability error analysis on the acquisition of experimental numbers. Repeatability error shown as equation (4) refers to repeating several measurements by the same person during the experiment, and comparing each measured result with the overall average. The relative error is of course as low as possible. If the repeatability error is too large, it means the measured data is floating and indicates that the data in this range is low in reference and the instrument test is unstable. Accuracy error refers to the relative error between the test result and the standard value of the gold sample as shown in equation (5). The method first selects several pure materials with known thermal diffusivity as golden samples, and compares the test results with the standard values of the test pieces to determine whether the instrument is accurate or not. If the accuracy error is too large, it means Instrument test results are not credible.

$$\varepsilon_{rep} = \frac{|\alpha - \alpha_{ave}|}{\alpha_{ave}} \times 100\% \tag{4}$$

$$\varepsilon_{std} = \frac{|\alpha - \alpha_{std}|}{\alpha_{std}} \times 100\% \tag{5}$$

$\alpha_{ave} = \frac{\sum_i^n \alpha_i}{n}$: Average of several measurements repeated. α_{std} : the standard value of the test piece.

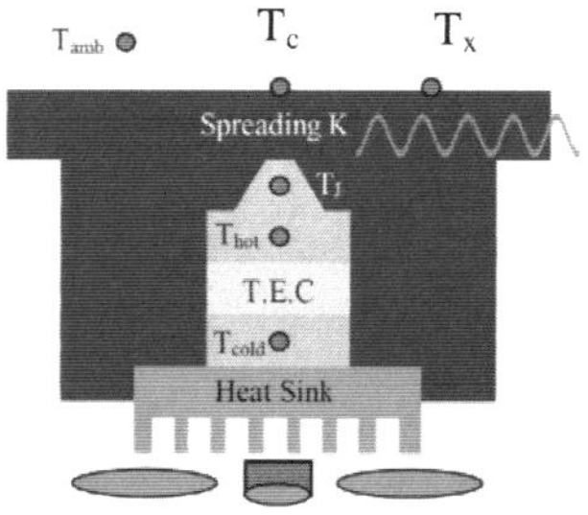

Figure 1. Schematic diagram of thermal diffusivity measurement

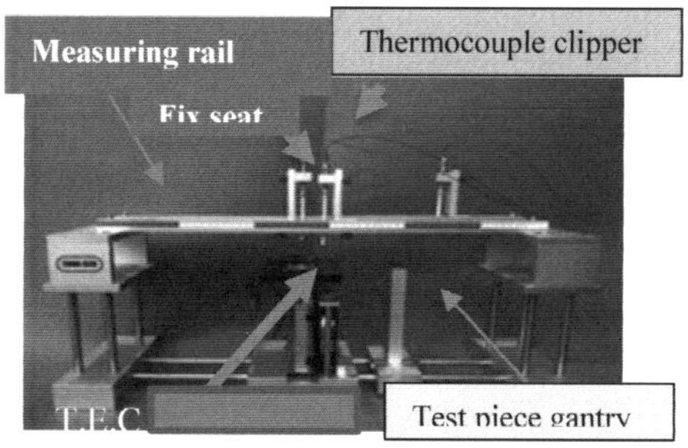

Figure 2. Heating platform for thermal diffusivity measurement system

3. Experiment results and discussion

3.1. The experiment verification of repeatability error

Table 2 shows the repeatability error $\varepsilon_{rep,i}$ is less than 5%. Results shows TDMI doing Well, the experimental data is highly reliable.

Table 2. Repeatability error analysis of Copper ($\alpha_{std,\,cu}$=1.17cm2/s)

L (mm)	W (mm)	T (mm)	τ=T/W	$\varepsilon_{cu,rep,1}$%	$\varepsilon_{cu,rep,2}$%	$\varepsilon_{cu,rep,3}$%
100	10	0.4	0.04	0%	0.84%	0%
100	20	0.4	0.02	0%	0.91%	0.91%
100	30	0.4	0.01	1.63%	0%	1.63%
100	40	0.4	0.01	3.38%	3.38%	3.38%
100	50	0.4	0.01	2.97%	3.44%	3.44%

3.2. The experiment verification of accuracy error

Table 3 shows the accuracy error analysis of copper in TDMI test. The diffusivity of pure copper $\alpha_{std,cu}$ is 1.17 cm^2/s. The average value of the experimental amount of 3 times is expressed by $\alpha_{cu,avg}$, when $\tau<0.02$, region is " A"; when $0.02\leq\tau\leq075$, regionis " B"; while $0.075<\tau$, region is " C". The red font dimension in the table are the accuracy error after partition correction. Table 2 shows that the accuracy error of copper is controlled within 10% after partition correction, as shown by the red character. Therefore, the thermal diffusivity correction is based on the measuring length 3 cm and the judgment distribution of one-dimensional and two-dimensional. The criterion for the judgment is the thickness-to-width ratio of the test sample as an index, the thickness of the test sample T must less than 5 mm or less.

(I) When $0.02 \leq \tau \leq 0.075$, it is "B"region, one-dimensional mode is adopted, $\alpha_{M,1D} = \frac{L^2}{2\Delta t(\ln\frac{M}{N})}$ (6)

(II) When $\tau<0.02$, it is "A"region, expressed as $\alpha_{M,1D\times2}$, $\alpha_{M,1D\times2} = \alpha_{M,1D} \times 2$ (7)

(III) When $\tau > 0.075$, it is "C" region, expressed as $\alpha_{M,1D\times0.5}$, $\alpha_{M,1D\times0.5} = \alpha_{M,1D} \times 0.5$ (8)

Table 3. Accuracy error analysis of pure copper diffusivity ($\alpha_{std,\ cu}$=1.17cm^2/s)

No	$\tau = \frac{T}{W}$	Zone	$\alpha_{cu,avg}$	1-D mode (B)	2-D mode (A)	0.5D mode (C)
				$\varepsilon_{cu,std,1D}$	$\varepsilon_{cu,std,1DX2}$	$\varepsilon_{cu,std,1DX0.5}$
1	0.04	B	1.198	2.4%	104.7%	48.803%
2	0.02	B	1.094	6.4%	87%	53.24%
3	0.013	A	0.615	47%	5.12%	73.71%
4	0.01	A	0.60	48%	2.564%	74.35%
5	0.008	A	0.5826	50%	0.41%	75.1%

3.3. Experimental results and analysis of graphite sheets

The main structure of the graphite sheet is a layer of PET material on the bottom layer, and then coated with graphene or graphite powder called graphite layer. Some of the manufacturers may have resin film on the graphite. The parallel thermal resistance is shown as in equation (9), If the thermal resistances expressed in the Fourier form, the total thermal conductivity $K_{eff,total}$ of the graphite sheet can simplify to equation (10). Table 4 shows the thermal diffusivity measurement results of the graphite sheets E and F and G. The thermal diffusivity of the graphite sheet for E, F, G are 1.43, 1.58 and 4.12 respectively. The total effective thermal conductivity of the graphite sheet E, F, G are 252, 328 and 590. The effective thermal conductivity of the graphite layer E is 452, F is 491 while for G is 1060.

$$\frac{1}{R_{th,total}} = \frac{1}{R_{th,PET}} + \frac{1}{R_{th,c}} + \frac{1}{R_{th,resin}} \tag{9}$$

$$k_{eff,total} = k_{PET} \times AF_{PET} + k_{eff,C} \times AF_C + k_{resin} \times AF_R \tag{10}$$

Table 4. Diffusivity measurement results of graphite sheets E, F and G (Test sample L=200mm, W=200mm)

sample	τ (H_{total}/W)	Zone	α_2 (cm^2/s)	graphite sheet K_{total} (W/m-k)	graphite layer $K_{eff,}$ (W/m-k)
sheet E	0.0015<0.02	A	1.43	$K_{eff,total,E}$=252	$K_{eff,total,C,F}$=452
sheet F	0.002<0.02	A	1.58	$K_{eff,total,F}$=328	$K_{eff,total,C,F}$=491
sheet G	0.0015<0.02	A	4.12	$K_{eff,total,G}$=590	$K_{eff,total,C,G}$=1060.9

3.4. Experimental results and analysis of the vapor chamber

The vapor chamber in this experiment has three kinds of VC-A, VC-B and VC-C, as shown in Fig. 3, Fig. 4 and Fig. 5, the dimensions are 90×90×2.35mm^3, 100×100×8mm^3 and 90× 90×0.85 mm^3. VC-A is a uniform shape vapour chamber with a thickness of 2.5mm. VC-B is an irregularly shaped vapour chamber with a thickness 8mm in the middle of the upper lid and the lower lid. VC-C is ultra-thin vapour chamber with thickness of 0.86mm.Table 5 shows the thermal diffusivity of "VC-A" for x-direction is α_x=1.99, while for Y-direction is α_y=1.77 and α_{total}=3.76. The thermal diffusivity of "B" for x-direction is α_x=0.55, while for Y-direction is α_y=0.54 and α_{total}=1.09. The thermal diffusivity of "C" for x-direction is α_x=0.48, while for Y-direction is α_y=0.37 and α_{total}=0.85. Convert the thermal diffusivity data to the thermal conductivity, that is equivalent to $K_{VC\text{-}A}$=19288, $K_{VC\text{-}B}$=4330 and $K_{VC\text{-}C}$=1614

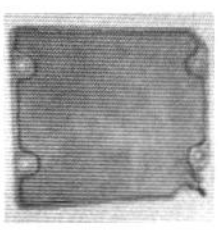

Figure 3. Vapor chamber VC-A

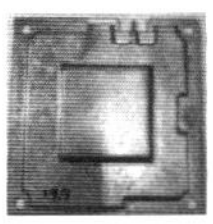

Figure 4. Vapor chamber VC-B

Figure 5. Vapor chamber VC-C

Table 5. Thermal diffusivity of VC-A ,VC-B and VC-C measured with two-dimensional mode

item	VC-A	VC-B	VC-C
ρ(g/cm^3)	$\rho_{ave,VC-A}$ =5.70	$\rho_{ave,VC-B}$ =4.43	$\rho_{ave,VC-C}$ =6.49
C(J/g·℃)	4.5	4.505	1.463
L×W×T (mm^3)	90×90×2.35	100×100×8	90*90*0.85
τ=T/W	τ=2.5/90=0.0261	τ=8/100=0.08	τ=0.85/90=0.0094
mode	Two-dimension Zone	Two-dimension Zone	Two-dimension Zone
α (cm^2/s)	α_{total}=3.76	α_{total}=1.085	α_{total}=0.85
$K_{VC,TDMI}$ (W/m-k)	K_{VC-A} =19288	K_{VC-B}= 4330.7	K_{VC-C}= 1614.1

4. Conclusion

4.1 According to the TDMI instrument, it is necessary to determine whether the experimental condition is a one-dimensional mode or a two-dimensional mode first. Generally, a width W is larger than 5cm defined a two-dimensional mode, and W=5 cm or smaller is a one-dimensional mode. The one-dimensional mode uses the aspect ratio as the criterion for the correction. The basis of the judgment is that the thickness-to-width ratio τ of the test piece is an index, and the material is applicable with thickness no larger than 5 mm

4.2 The K value (vertical phase) of the natural graphite measured by D5470 in the general literature is about 25 (W/m-k), and the k value in the horizontal direction is between 500 and 1200 (W/m-k). The effective thermal conductivity of three typical graphite layers have been measured by TDMI which are 452(W/m-k), 491(W/m-k) and 1060.9(W/m-k) respectively. The TDMI, designed with the Angstrom principle provides a relatively inexpensive, fast and simple measuring method of the effective thermal conductivity of graphite sheets and graphite layers.

4.3 TDMI can also calculate its thermal conductivity after two-dimensional correction. Since there is no standard thermal diffusivity value of the vapor chamber, it can only use as a reference. In this study, three pieces of vapor chamber VC-A, VC-B and VC-C were measured which are 20,000(W/m-k), 4000(W/m-k) and 1600(W/m-k) respectively.

References

[1] J.E. Parrott and A.D. Stuckes, "Thermal Conductivity of Solids", Pion Limited, London, 1975.

[2] ASTM D5470-01, Standard Test Methods for Thermal Transmission Properties of Thin Thermally Conductive Solid Electrical Insulation Materials, American Society for Testing and Materials, West Conshohocken, PA.

[3] K. L. Wray and T. J. Connolly, "Thermal conductivity of clear fused silica at high temperatures", Journal of Applied Physics, vol. 30, pp. 1702-1705, 1959.

[4] V. Mirkovich, "Comparative method and choice of standards for thermal conductivity determinations", Journal of the American Ceramic Society, vol. 48, pp. 387-391, 1965.

[5] D. Hughes and F. Sawin, "Thermal conductivity of dielectric solids at high pressure", Physical Review, vol. 161, p. 861, 1967.

[6] A. Tomokiyo and T. Okada, "Determination of thermal diffusivity by the temperature wave method", Japanese Journal of Applied Physics, vol. 7, p. 128, 1968.

[7] Chen-I Chao, Wei-Keng Lin,* Shao-Wen Chen, & Han-Chou Yao, "FEASIBILITY OF THE ÅNGSTRÖM METHOD IN PERFORMING THE MEASUREMENT OF THERMAL CONDUCTIVITY IN VAPOR CHAMBERS", Heat Transfer Research 47(7), 617–632, 1064-2285/16/ $35.00 © 2016 by Begell House, Inc. (2016) , (MOST 104-ET-E-007-005-ET)

IOP Conf. Series: Materials Science and Engineering **600** (2019) 012024 doi:10.1088/1757-899X/600/1/012024

DC-link voltage feedforwarded interpolation error compensation method for field weakening operation region of look-up table based PMSM drive

Jung-Hyo Lee[1,*]

[1,*]Kunsan National University, Electrical Engineering Department, Gunsan 54150, Korea

jhlee82@kunsan.ac.kr

Abstract. This paper proposes field weakening control method of the look-up table based permanent magnet synchronous machine (PMSM) with interpolation error compensation. The look-up table based control method has robust control characteristic than other control methods which use linear controller for current reference generation. However, it is impossible to store all current references under all circumstances for torque commands. General look-up table based control method using two input parameters, in order to mitigate the effect of discretely stored data, two-dimension-interpolation method (2D-Interpolation) is used for linear interpolating between discontinuous data. However, due to the current trajectories of PMSM are generally ellipsoidal, the error is occurred between linearly interpolated and controllable current references. This paper proposes a method to compensate this interpolation error using feedforward controller for fast compensation. The improvement of proposed method is verified by experiment and simulation.

1. Introduction

Recently, Permanent Magnet Synchronous Motor (PMSM) drive system has been adopted frequently for vehicles in order to improve efficiency and convenience. Generally, in the case of the PMSM used in a vehicle, two dimensions look-up table (2D-LUT) based current command control is widely used.[1]-[6] Among these control methods, flux-torque 2D-LUT based control method is most generally used due to the reflection of DC-link voltage variation which shown as Fig. 1.[1]

In this controller, the current data in the 2D-LUT is stored according to the respective memory addresses corresponding to flux-torque. Since the memory address is a discontinuous value, only current reference data for a specific flux and torque can be stored in the memory. Therefore, in the case of the PMSM torque control method using the 2D-LUT, it is inevitably impossible to store the appropriate current data in the entire driving region. In order to solve this problem, the two-dimension-interpolation (2D-Interpolation) method has been used to properly interpolate with input data that has not been previously stored in the memory.[7]-[9] This 2D-Interpolation makes discreet look-up table outputs to linear outputs according to two continuous input parameters which do not define at the specific torque-flux points of the look-up table. For instance, if the input parameters inserted in look-up table, related outputs are generated from the memory, and then, using these outputs for linear interpolation called first order newton's interpolation method in order to calculate approximate output for input parameters.

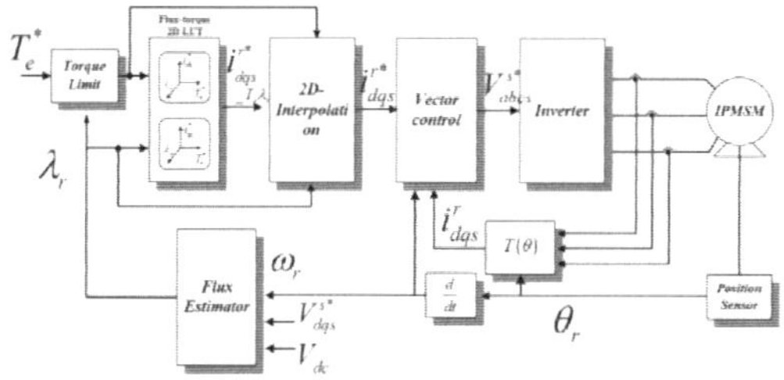

Figure 1. Flux-torque look-up table based PMSM control method [1].

This 2D-Interpolation method used in PMSM control is effective if the number of stored data is enough to ignore the interpolation error, however, if the number is insufficient, interpolation error can not be ignored because all PMSM current trajectories are not linear but ellipsoidal. Moreover, in automotive application, in order to fulfill the international standard ISO26262 which purpose is to retain driver safety, PMSM drive should have many fault diagnosis for retaining system safety and AUTOSAR which needs lots of memory for operating software.

To solve this problem, very few researches have been proposed to optimal use of memory. In [10], use the curve fitting method for reducing the memory, however, it does not suggest the solution for the error between the interpolated curve fitting output and optimal output. Constant torque control method for PMSM using any tables is shown on [11], however, this method needs exact parameters for PMSM, moreover, if the target PMSM does not have enough saliency, effectiveness is limited. As aforementioned, current trajectories for PMSM are ellipsoidal, the interpolation method used in this application needs second order interpolation method such as second order Lagrange or Spline method with parameter modification for each operating condition. [12] However, it is seldom used in motor control area because of heavy calculation burden to DSP.

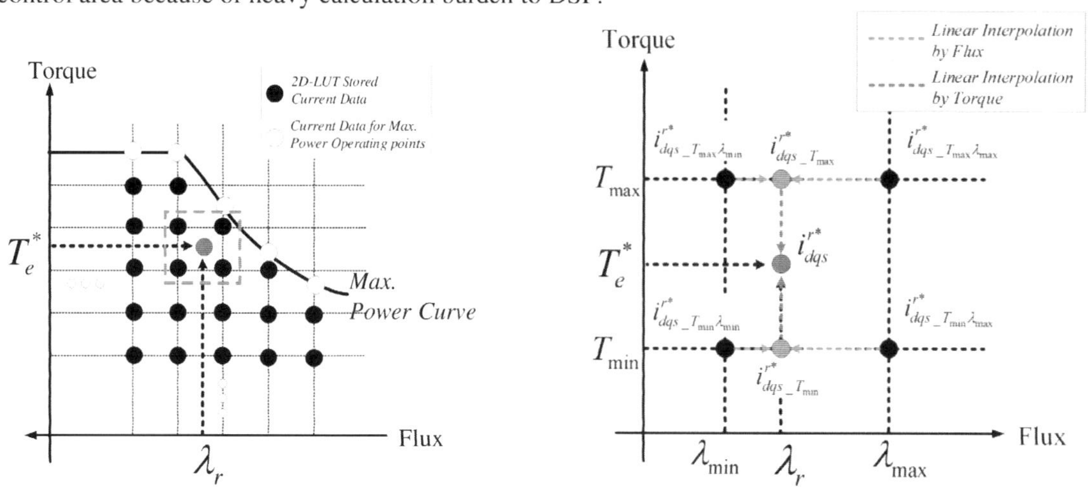

(a) (b)

Figure 2. 2D-Interpoation procedure for current references output from 2D-LUT. (a) Stored data in 2D-LUT accordance with each speed and torque. (b) 2D-Interpolation procedure for specific speed and torque reference input.

This paper proposes a novel control method to reduce this interpolation error. First, we analyze the cause of interpolation error and defines the problem characteristic. And then, compensation method for interpolation error with DC-link voltage feedforward will be illustrated.

2. Interpolation error of 2D-Interpolation technique for PMSM drive

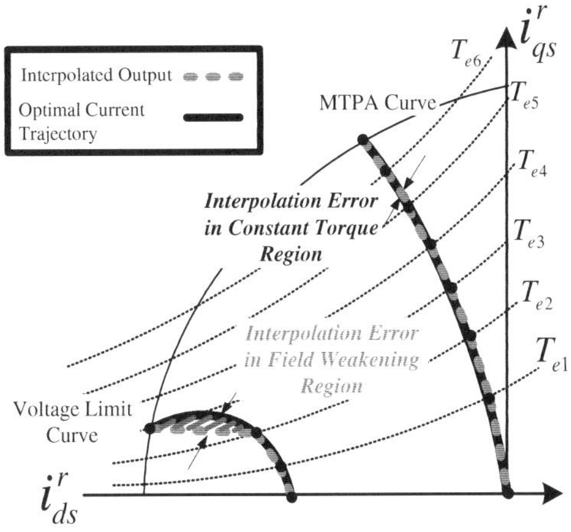

Figure 3. Interpolated current trajectories of PMSM with the current table data which has six torque input data.

2.1. 2D-Interpolation Technique

The 2D-Interpolation is a linear interpolation method for two input parameters. As aforementioned, 2D-LUT data on certain flux and torque have four different values: outputs on the ceil and floor of flux input (λ_{max} , λ_{min}); outputs on the ceil and floor of torque input (T_{max}, T_{min}) as shown on Fig. 2. To generate interpolated output, the following linear interpolation formula is applied to the input parameters three times as shown on Fig. 2(b).

$$f(x)=f(x_1)+\left(\frac{f(x_2)-f(x_1)}{x_2-x_1}\right)\cdot(x-x_1) \tag{1}$$

For instance, the case of linear interpolation by speed on T_{min}, $x_2$and x_1 are ceil and floor of speed input (λ_{max} , λ_{min}), respectively, $f(x_2)$ and $f(x_1)$ are the current reference data on T_{min} correspond to λ_{max} , λ_{min},the notations of Fig. 2(b) are $i^{r^*}_{dqs\_T_{min}\lambda_{max}}$, $i^{r^*}_{dqs\_T_{min}\lambda_{min}}$, respectively. x is the input of current speed (λ_r) and $f(x)$ is the interpolated output on T_{min} ($i^{r^*}_{dqs\_T_{min}}$).

To reduce the processing burden of microprocessor, division for variable has to be avoided as possible. This paper uses fixed unit for that as same as [9]. In this case, the current references from look-up table's are also established accordance with this unit value. Substitute the variable to fixed unit value, (1) can be changed as below.

$$f(x)=f(x_1)+\left(\frac{f(x_2)-f(x_1)}{x_{unit}}\right)\cdot(x-x_1) \tag{2}$$

where, x_{unit} is the fixed unit value of input parameter's deviation.

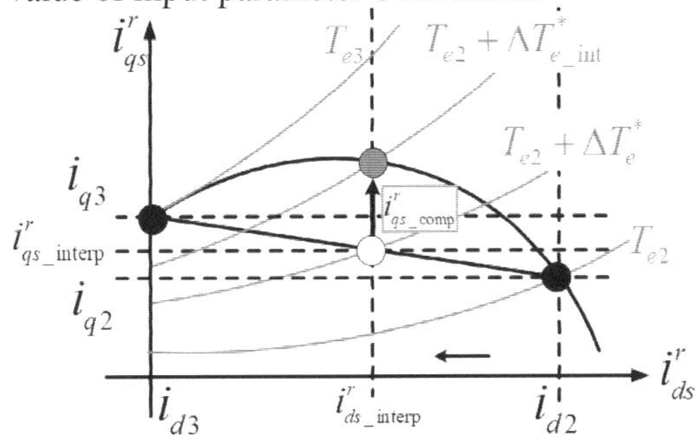

Figure 4. Magnified 2D-Interpolation error in field weakening operation region of Fig. 3.

Note that the flux is inverse value of the speed. If the speed is increasing, then, the flux is decreasing as shown on Fig. 2(a). Therefore, the calculated flux should be properly limited to avoid divergence.

2.2. 2D-Interpolation Error in Look-Up Table Based PMSM Control

Fig. 3 shows the interpolated current trajectories using only six torque inputs data. As shown on the figure, with 2D-Interpolation method, two current reference points stored in the memory are linearly connected while the operating condition that stored data is absent. Although it can reduce the error better than using only look-up table outputs, generated power is decreased than capable maximum power. In addition, the effect of interpolation error is hardly existed in constant torque operation region due to the sufficient current data for torque input. In contrast, the effect of interpolation error is enlarged in field weakening operation region because the current data is reduced in high speed operation.

Fig. 4 shows the magnified interpolation error described in Fig. 3. To calculate this 2D-Interpolation error, firstly, fundamental PMSM model equations are shown as below.

$$\begin{aligned} v_{ds}^r &= R_s i_{ds}^r + L_d p i_{ds}^r - \omega_r L_q i_{qs}^r \\ v_{qs}^r &= R_s i_{qs}^r + L_q p i_{qs}^r + \omega_r L_d i_{ds}^r + \omega_r \phi_f \end{aligned} \quad . \tag{3}$$

$$T_e = 3P/4\{\phi_f i_{qs}^r - (L_q - L_d) i_{ds}^r i_{qs}^r\} \tag{4}$$

$$\begin{aligned} V_{\max} &= \omega_r \sqrt{(L_d i_{ds}^r + \phi_f)^2 + (L_q i_{qs}^r)^2} \\ V_{\max} &= \frac{V_{dc}}{\sqrt{3}} \end{aligned} \quad . \tag{5}$$

where, v_{ds}^r, v_{qs}^r are d-/q-axis voltages of PMSM respectively, T_e is torque, $V_{\max}$ is voltage restriction of PMSM, ω_r is angular speed, R_s is phase resistance, L_d, L_q are d-/q-axis inductances, respectively, ϕ_f is permanent magnet flux, P is number of pole, V_{dc} is DC-link voltage.

As shown on the figure, the interpolation error does not affect d-axis current component ($i_{ds\_interp}^r$), however, it affects the q-axis current component ($i_{qs\_interp}^r$). This interpolated q-axis current reference can be expressed as (6) from (1).

$$\begin{aligned} & i_{qs\_interp}^r (T_{e2} + \Delta T_e^*) = i_{q3} + \frac{i_{q3} - i_{q2}}{T_{e3} - T_{e2}} \Delta T_e^* \\ & (\Delta T_e^* = T_e^* - T_{e2}) \end{aligned} \tag{6}$$

With (6), the back-EMF voltage magnitude by interpolated current reference can be expressed as (7).

$$\begin{aligned} V_{mag} &= \omega_r \sqrt{(L_d i_{ds\_interp}^r (T_{e2} + \Delta T_e^*) + \phi_f)^2 + (L_q i_{qs\_interp}^r)^2} \\ V_{mag} &< V_{\max} \end{aligned} \tag{7}$$

If the interpolated d-axis current according to the varied torque with (7), the magnitude of the maximum torque that can be generated by this d-axis current can be obtained through the intersection of the voltage restriction curve and the torque curve. The equation is expressed by a quadratic equation as shown in (8) which has four roots that are classified as two imaginary roots, and two real roots. Among these two real roots, effective d-axis current is the magnitude which has the minimum value of them.

IOP Conf. Series: Materials Science and Engineering **600** (2019) 012024 doi:10.1088/1757-899X/600/1/012024

$$\begin{aligned}&-(L_d-L_q)^2L_q^{\,2}{i^r_{ds\_interp}}^4\\&+\left\{-2L_d^{\,2}(L_d-L_q)-2\phi_fL_d(L_d-L_q)^2\right\}{i^r_{ds\_interp}}^3\\&+\left\{-2L_d^{\,2}\phi_f^{\,2}-4\phi_fL_d(L_d-L_q)-(L_d-L_q)^2\phi_f^{\,2}+(L_d-L_q)^2\left(\frac{V_{max}}{\omega_r}\right)^2\right\}{i^r_{ds\_interp}}^2\\&+\left\{-2\phi_f^{\,3}L_d-2(L_d-L_q)\phi_f^{\,2}+2(L_d-L_q)^2\left(\frac{V_{max}}{\omega_r}\right)^2\right\}i^r_{ds\_interp}\\&+\left\{\phi_f^{\,2}\left(\frac{V_{max}}{\omega_r}\right)^2-\left(\frac{L_q(T_{e2}+\Delta T_{e\_int})}{P_n}\right)^2-\phi_f^{\,4}\right\}=0\end{aligned} \tag{8}$$

The compensation value of q-axis current can be obtained by substituting d-axis current value from (8) to (4)

$$i^r_{qs\_comp}=\frac{4}{3P}\frac{T_{e2}+\Delta T^*_{e\_int}}{(\phi_f+(L_d-L_q)i^r_{ds\_interp})}-i^r_{qs\_interp} \tag{9}$$

From (9), increased torque output by compensated q-axis current ($\Delta T^*_{e\_int}$) is always higher than increased torque output by interpolated q-axis current (ΔT^*_e) as shown on Fig. 4.
As a result, it proves that the torque error is existed between interpolated output and optimal output, in addition, generated torque by interpolated output is always smaller than the generated torque by optimal output.

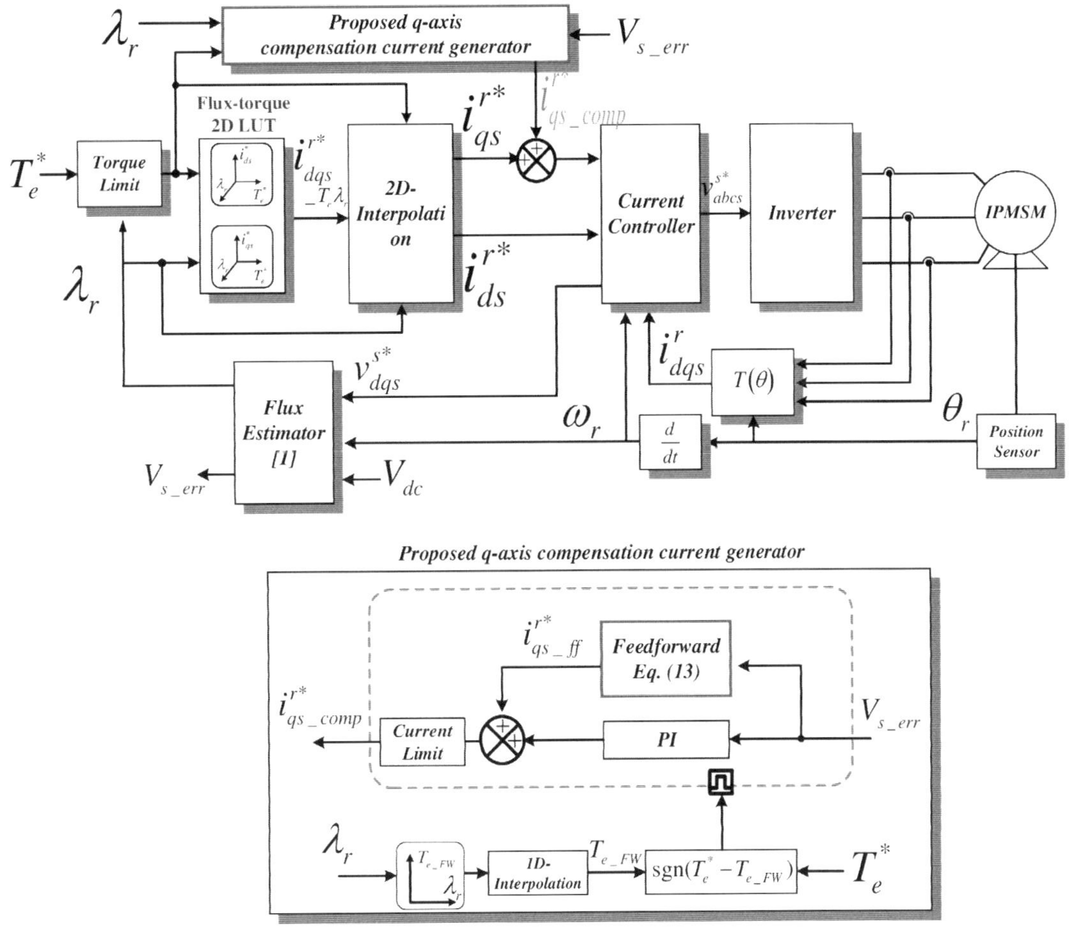

Figure 5. Overall proposed control block for q-axis current compenstion.

3. Interpolation Error Compensation Method with DC-link voltage feedforward Controller

Generally, in field weakening operation region, since the magnitude of the current is not large, the magnitude of the back-EMF voltage has a major influence on the components in permanent magnet flux. Therefore, the variation in the torque in this region is largely affected due to the variation in the q-axis current, and the voltage restriction is greatly affected by the speed and the d-axis current. Among them, the speed is not a direct control target of the electric motor but a restriction condition that is influenced by the machine-driving environment. Therefore, in order to compensate reduced output caused by the interpolation, q-axis current must be compensated so as to raise the torque.

The reason of occurring this torque error is the back-EMF magnitude of interpolated q-axis current is smaller than the voltage limit, which is determined by DC-link voltage. Therefore, compensated q-axis current can be obtained from the voltage error between the current back-EMF magnitude and voltage limit calculated by DC-link voltage. From this error, PI controller in this paper deducts compensated q-axis current.

$$i^r_{qs\_comp} = \frac{sK_p + K_i}{s}(V_{max} - V_{mag}) \tag{10}$$

However, as previously mentioned, all current trajectories for PMSM operation is ellipsoidal, PI controller is insufficient to calculate compensation current. As widely known, PI controller is effective in fixed reference, however, if the reference is ellipsoidal, the final output value has always error from delayed response.

The easiest way to respond suitably for ellipsoidal reference, using PID controller to increase the pole of the controller's characteristic equation or adapting numerous higher gain of PI controller for reducing this final output error. However, D controller needs suitable filter for avoiding divergent output, higher gained PI controller can be very unstable operation for disturbance.

In this paper, feedforward controller is added to improve the dynamic of PI controller. To obtain the feedforward compensation current, voltage error between the back-EMF and the voltage limit can be obtained as (11) with (7).

$$\begin{aligned} V_{s\_err} &= V_{max} - V_{mag} \\ &= \frac{V_{dc}}{\sqrt{3}} - \sqrt{(\omega_r L_q i^r_{qs\_linear})^2 + (\omega_r L_d i^r_{ds\_linear} + \phi_f)^2} \end{aligned} \tag{11}$$

As shown on Fig. 4, because interpolated d-axis current reference generates proper current for the voltage limit, (11) can be changed to (12).

$$\begin{aligned} V_{s\_err} &= V_{max} - V_{mag} \\ &= \frac{V_{dc}}{\sqrt{3}} - \omega_r L_q \left| i^r_{qs\_linear} \right| \end{aligned} \tag{12}$$

To make this voltage error to zero, compensation feedforward q-axis current can be established as (13).

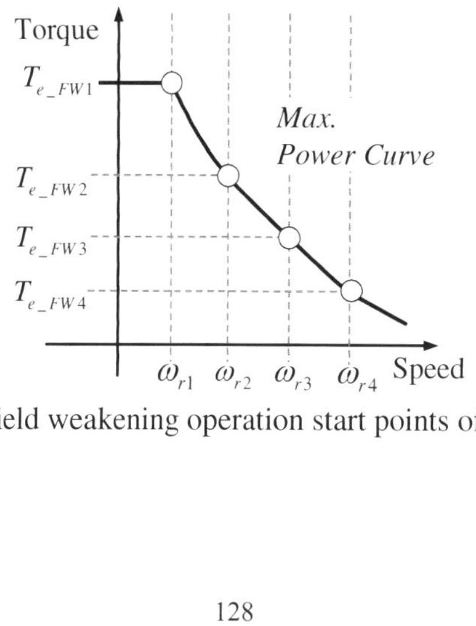

Figure 6. Field weakening operation start points of each speed

$$i^{r}_{qs\_ff} = \frac{V_{s\_err}}{\omega_r L_q} \tag{13}$$

With (13), proposed overall control block for q-axis compensation current is established as Fig. 5.
As aforementioned, interpolation error is hardly existed in constant torque operation region. Moreover, the back-EMF in this operation region has its own value at each operating point, compensating interpolation error in this operation region needs another back-EMF look-up table data according to each flux and torque. Therefore, in this paper, compensation block is deactivated while the PMSM is operating in constant torque operation region.

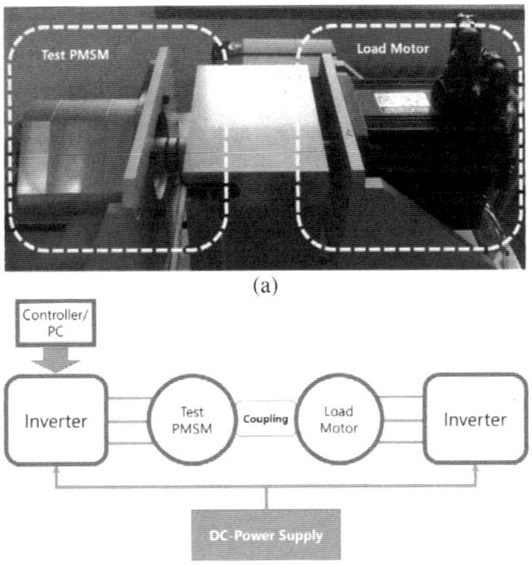

Figure 7. Experimental Msetup (a) Picture of the setup (b) Composition of the setup

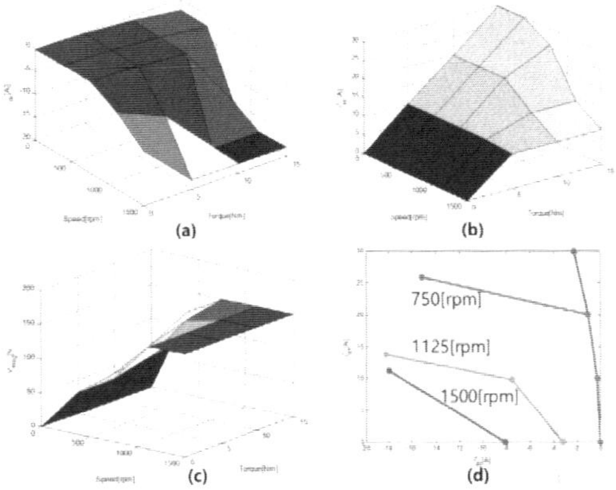

Figure 8. Current maps using in experiment (a) d-axis current map (b) q-axis current map (c) Back-EMF magnitude (d) dq-axis current maps plotting by dq-axis currents coordinates.

To reduce the stored memory for voltage magnitude, field weakening operation start points of each speed are obtained experimentally instead of obtaining voltage magnitudes all over the operation region. Fig. 6 shows the stored data of field weakening start points. As shown on the figure, as speed is increasing, generated torque is reducing to constant power generation. In addition, the error between practical torque-speed curve and linearly interpolated curve is hardly existed, the field weakening start torque, which is generated from flux-torque LUT of Fig. 5 can determine enable or disable compensation block. Instead of using speed input, estimated flux is used for reflecting input voltage variation. Converting method from speed to flux is simply defined as (14).

$$\lambda_r = V_{\max} / \omega_r \tag{14}$$

And then, activation condition of q-axis current compensation can be determined by following equation.

$$\begin{aligned} &if\,(T_e^{*} - T_{e\_FW} > 0) : \operatorname{sgn}(T_r^{*} - T_{e\_FW}) = 1 \\ &if\,(T_e^{*} - T_{e\_FW} \le 0) : \operatorname{sgn}(T_r^{*} - T_{e\_FW}) = 0 \end{aligned} \tag{15}$$

With this enable/disable control block, the proposed compensation block is only activated when the effect of compensation is maximized.

IOP Conf. Series: Materials Science and Engineering **600** (2019) 012024 doi:10.1088/1757-899X/600/1/012024

4. Experiment

To verify the proposed control algorithm, this paper uses experiment setup as Fig. 7. The test motor parameters are described in Table 1. Controller of inverter uses DSP, which is TMS320F28335 from Texas Instrument corporation.

Although this DSP has sufficient memory for storing current 2D-LUT data, for the worse environment configuration, the current map is constructed using very little memory compared to the conventional method.

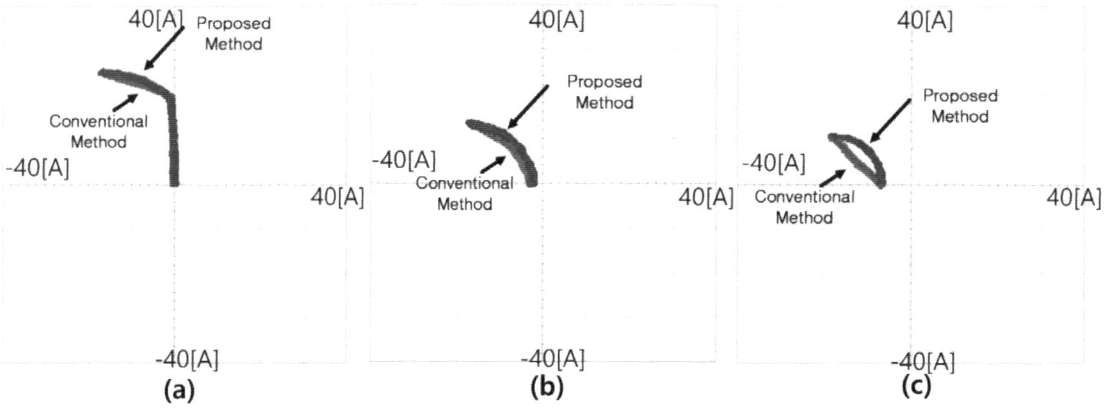

Figure 9. Comparison experimental results of the conventional and the proposed method with or without 2D-Interpolation compensation (a) 750[rpm] operation (b) 1150[rpm] operation (c) 1500[rpm] operation

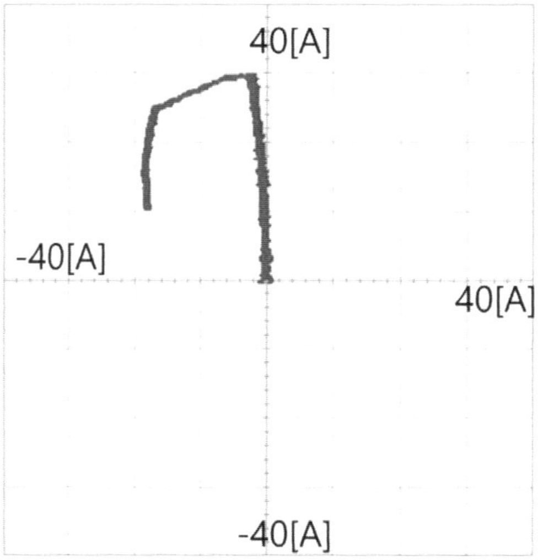

Figure 10. Experimental result of the maximum power control with proposed control method

The dq-axis current map for each speed-torque was measured through experiments by referring [6]. The results are shown in Fig. 8. In this paper, the experimental current map is stored at 1/3 of the rated torque and at 1/4 of the maximum driving speed. From this obtained speed-torque current map, it is transformed to flux-torque map with (14). As a result, it can be seen that only one torque data appears in the dq-axis current value for the torque command change at the maximum speed except for the dq-axis current value during the zero torque control. This is because the inductance expressed by the nominal value in the actual motor and the permanent magnet flux decreased due to the influence of saturation in the actual experiment.

Fig. 9 shows the comparison experimental results with or without proposed compensation method. It can be seen from the figure that the output is improved compared with the conventional method, and that it is controlled stably even in the speed region where the transition from the constant torque region to the weak field region is performed.
Fig. 10 shows the result of testing the maximum output control by increasing the speed of the load motor after setting the torque command to the maximum. As can be seen from the figure, it can be seen that the maximum output control is performed stably even if the proposed controller is added.

5. Conclusion

This paper proposes interpolation error compensation method for general flux-torque 2D-LUT based PMSM control method. To achieve this, firstly calculate the error between the back-EMF and the voltage limit in field weakening region, and then, obtaining q-axis compensation current from the controller, which input is this voltage error. To improve compensation control dynamic for ellipsoidal reference, feedforward controller, which is based on DC-link voltage, is attached on general PI controller.
With the proposed compensation blocks, generated power is enhanced than the power using only 2D-Interpolation method, although very restricted number of memory is allowed for flux-torque 2D-LUT. Proposed control method is verified by the experiment with test setup.

Table 1. Test motor parameters.

Pole	8
Phase resistance[mΩ]	20
d-axis inductance[mH]	2.03
q-axis inductance[mH]	2.13
Permanent magnet flux[Wb]	0.1439
Rated DC-link voltage[V]	48
Rated Torque[Nm]	15
Rated Speed[rpm]	200
Max. Speed[rpm]	1500
Max. Phase Current[A]	30

Acknowledgments

This research was supported by Basic Science Research Program through the National Research Foundation of Korea(NRF) funded by the Ministry of Education(NRF-2018R1C1B6008895).
This research was supported by Basic Science Research Program through the National Research Foundation of Korea(NRF) funded by the Ministry of Education(NRF- 2016R1A6A1A03013567).

References

[1] Y. Kusaka, E, Yamada, and Y. Kawabata, "Method and Apparatus for Driving and Controlling Synchronous Motor using Permanent Magnets as its Field System," US Patent 55169995, Oct, 1996.
[2] JaeHyuk Lee, JungHyo Lee, JinHo Park, ChungYuen Won, "Field-weakening strategy in condition of DC-link voltage variation using on electric vehicle of IPMSM" *IEEE-Conf. Electrical Machines and Systems* (ICEMS), pp. 1-6, Aug. 2011.
[3] Yang Nanfang, Luo Guangzhao, Liu Weiguo, Wang Kang, "Interior permanent magnet synchronous motor control for electric vehicle using look-up table" *Inter. Conf. on Elec. Mach. Sys.(ICEMS)*,pp.1015-1019,Jun.2012.
[4] Tae-Suk Kwon, Gi-Young Choi, Mu-Shin Kwak and Seung-Ki Sul "Novel Flux-Weakening Control of an IPMSM for Quasi-Six-Step Operation, " *IEEE-Trans. Ind. Appl.*, vol 44, no. 6 pp. 1722-1731, Nov. 2008.
[5] Bing Cheng, Tesch, T.R., "Torque Feedforward Control Technique for Permanent-Magnet Synchronous Motors," *IEEE-Trans. Ind. Elect.*, vol 57, no. 3, pp. 969-974, Mar. 2010.

[6] Jung-Hyo Lee, Chung-Yuen Won, Byoung-Kuk Lee, Hyun-Bae Kim, Jei-Hoon Baek, Kyu-Bum Han, U-In Chung, "IPMSM Torque Control Method Considering DC-link Voltage Variation and Friction Torque for EV/HEV applications," *IEEE-Conf., Vehicle Power and Propulsion Conference (VPPC),* pp.1063–1069, 2012.

[7] Marco Tursini, Enzo Chiricozzi, Roberto Petrella, "Feedforward Flux-Weakening Control of Surface-Mounted Permanent-Magnet Synchronous Motors Accounting for Resistive Voltage Drop", *IEEE-Trans. on Ind. Elec.*, vol 57, no. 1 pp. 440-448, Jan. 2010.

[8] Jin-Ho Park, Jung-Hyo Lee, Jae-Hyuk Lee, Chung-Yuen Won, "Current Control Method of IPMSM in Constant Power Region for HEV", *IEEE-Conf., Inter. Conf. on Elec. Mach. Sys.(ICEMS)*,pp.1015-1019, Aug.2011.

[9] Robert U. Lenke, Rik W. De Doncker, Mu-Shin Kwak, Tae-Suk Kwon, Seung-Ki Sul,"Field Weakening Control of Interior Permanent Magnet Machine using Improved Current Interpolation Technique" *IEEE-Conf. Power Electronics Specialists Conference (PESC),pp. 1-5,* 2006.

[10] Jorge G. Cintron-Rivera, Shanelle N. Foster, Carlos A. Nino-Baron, Elias G. Strangas, "High performance controllers for Interior Permanent Magnet Synchronous Machines using look-up tables and curve-fitting methods", *IEEE-Conf. International Electric Machines & Drives Conference (IEMDC),* pp. 268 - 275, 2013.

[11] Young-Doo Yoon, Wook-Jin Lee, Seung-Ki Sul, "New flux weakening control for high saliency interior permanent magnet synchronous machine without any tables",*IEEE-Conf. European Conference on Power Electronics and Applications (ECPEA)*, pp. 1 - 7, 2007.

[12] Shoudao Huang, Ziqiang Chen, Keyuan Huang, Jian Gao, "Maximum Torque Per Ampere and Flux-weakening Control for PMSM Based on Curve Fitting", *IEEE-Conf. Vehicle Power and Propulsion Conference (VPPC),* pp. 1 - 5, 2010.

IOP Conf. Series: Materials Science and Engineering **600** (2019) 012025 doi:10.1088/1757-899X/600/1/012025

ESD protection circuit for V-band RF applications in a 65nm CMOS technology

Allenn C. Lowaton[1], Lemuel Fil L. Bañez[2]

Microelectronics Laboratory, Electrical Engineering and Technology Department, Mindanao State University – Iligan Institute of Technology, Iligan City, Philippines

[1]allenn.lowaton@g.msuiit.edu.ph; [2]lemuelfilbanez@g.msuiit.edu.ph

Abstract. To apply radio frequency circuits, nanoscale CMOS technologies are greatly used but this consequence to a thinner gate oxide and silicided drain/source. With this, the electrostatic discharge (ESD) robustness of RF circuits will be much more degraded. Therefore, there is a need for an ESD protection circuit design in RF circuits against ESD damages. In this paper, the proposed ESD protection circuit is presented with concept based silicon-controlled rectifier (SCR) device, two diodes, PMOS transistor and an inductor to efficiently provide protection for radio-frequency (RF) circuits from ESD damages in nanoscale CMOS process. The concept based SCR device is aided with an inductor to provide efficient ESD path discharge to the SCR device. Moreover, the inductor is used to resonate out the parasitic capacitance generated by the ESD protection circuit at the preferred frequency. In addition, the proposed ESD protection circuit is implemented to a low noise amplifier (LNA) at 60-GHz frequency within an efficient area. Furthermore, the ESD protection circuit is simulated to achieve over 2-kV human body model ESD robustness with good RF performances on LNA in 65-nm CMOS technology.

1. Introduction

The increasing trend of wireless communication has been promising since the emergence of fast and compact smart phones in the global market. However, the use of these devices expressed some issues on effects of electrostatic discharge (ESD) on the devices performance and reliability. As the current technology advances, more concerns are raised on the association between ESD and decreasing size of devices [1]. In modern integrated circuits(IC), the problem of ESD has become the principal reliability issue as ICs are manufactured from fabrication to the use of consumers. ESD zapping can happen at any moment at different places and the trouble of ESD damages increases as thinner gate oxides and smaller diffusion junction arises [1].According to ESD Association, 25% of all electronics damaged with causes that were not known are designated to be an ESD related problem and around $5 billion (USD) per

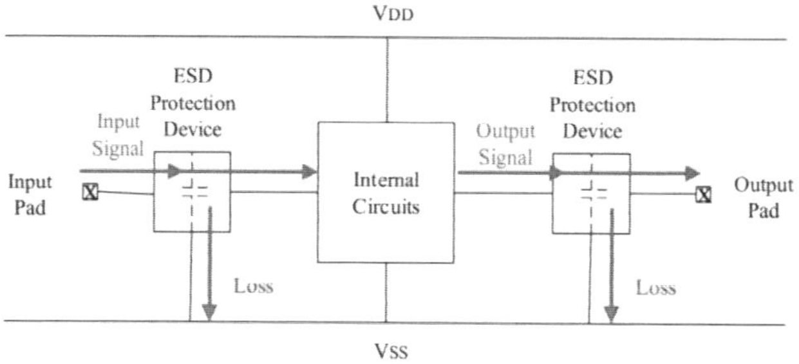

Figure 1. Signal Loss at I/O Pads of IC with ESD Protection Devices

year is the cost of ESD damage to the Electronics Industry [2]. The failures of CMOS are attributed to ESD damage which cost the IC industry billions of dollars [3], therefore the existence of ESD problems can no longer be neglected. To avoid electrostatic discharge, precautions were made and ICs should be capable to withstand ESD events for it to become reliable. Identifying the ESD susceptible devices and protect them with an ESD protection circuit should be included in the design process. Furthermore, designers should test the circuits with ESD protection to ensure its reliability under ESD stresses [1]. As stated in the International Technology Roadmap for Semiconductors (ITRS), the design of on-chip ESD protection circuit to radio frequency (RF) circuits rises as a primary challenge in its reliability and performance. Since RF ICs operates at high frequencies, these circuits are highly sensitive to any parasitic effects created by ESD protection structures which always affect the RF ICs performance [4]. One of the most important design considerations for RF circuits is the parasitic capacitance which comes from the addition ESD protection devices. Using large device dimen- sions in conventional ESD protection circuit will have parasitic capacitance that is too large to be tolerated for RF circuit applications [5]. Figure 1 shows the effect of parasitic capacitance on ESD protection devices. The parasitic capacitance causes signal loss from the input/output pad to ground. Furthermore, it changes the input matching condition of the device which results to the deterioration of RF performance [5]. To lessen the reduction performance degradation caused by the addition of ESD protection circuits, some design techniques have been developed to minimize its parasitic capacitance [5]. The ESD protection circuit with smaller parasitic capacitance can be easily co-designed or combined with RF circuits [6]-[7].

In this study, a new design approach is proposed for ESD protection circuit which will improve the ESD robustness of LNA under ESD events. Furthermore, a two-stage cascode topology is used for the LNA circuit which achieves high gain and good impedance matching. Both the proposed designs utilize the Synopsys IC design tools.

2. System architecture

The design of ESD Protection circuit is composed of two blocks, the SCR-Based ESD protection block and the Power- rail ESD clamp block. Furthermore, to evaluate the performance of the ESD protection circuit, a low noise amplifier block is used as the device under test. As shown in Fig. 2, the SCR-Based ESD protection serves as the main block of ESD protection circuit since it is connected with the input port while the power-rail ESD clamp circuit provides protection of the device under test from ESD events that flows between supply voltage and ground. Lastly, the low noise amplifier is the device under test that is ESD Protection Circuit Device under Test being protected using the proposed ESD protection design. The proposed system architecture is shown in Fig. 2.

2.1. SCR concept based ESD protection

Among the ESD protection devices, such as the diode, MOS, BJT or field-oxide device, silicon controlled rectifier (SCR) device has been revealed to be useful for RF ESD protection design due to its high ESD robustness within a small layout area and low parasitic capacitance [8]. Aside from that, the SCR device usually has a holding voltage of ~1.5 V in the bulk CMOS processes [9]. Furthermore, as the supply voltage (VDD) decreases in a 65-nm CMOS process, the SCR can be safely used without latchup danger [8].Silicon Controlled Rectifier has a pnpn structure with two parasitic transistors. Fig. 3 shows the cross sectional view as well as the equivalent ciruit of SCR device.The equivalent circuit of the SCR consists of a PNP BJT (Qnpn) and an NPN BJT (Qnpn), as shown in Fig. 3(b). The (Qnpn) is formed by the N-well, P-well, and N+ [11], and the (Qpnp) is formed by the P+, N-well, and P-well. By turning on the parasitic npn bipolar transistor due to avalanche breakdown generated current, the SCR is triggered. Furthermore, SCR has the capability to switch from high impedance state to low impedance state since the feedback mechanism keeps both the bipolar transistors on and therefore [1].The SCR path between RFin and Vss consists of P+, N- well, P-well, and N+. Moreover, the parasitic diode path between RFIN and VDD consists of P+ and N-well/N+. When ESD zapping occurs from RFin to Vss, the positive-feedback regenerative mechanism of Qpnp and Qnpn of the SCR will be highly conductive making SCR very robust against ESD stresses [9]. Due to almost ideal conductivity modulation of SCR, it can conduct very high current with very low on resistance. SCR can efficiently discharge ESD current in both directions since it provides full voltage swing. Also, SCR has low parasitic capacitance, higher secondary breakdown current, It2, and smaller area which offer a favorable choice for ESD protection designs. However, the drawback of using SCR must be solved like higher first breakdown voltage and lower holding voltage (causes latch up) [8].There have been various modified SCR designs that were developed to solve the disadvantages of SCR, making SCR the most preferred ESD protection device [1]. The trigger signal can be sent to enhance the turn-on speed to reduce the trigger voltage of an SCR device. Some design techniques have been reported to enhance the turn-on efficiency of SCR devices [9]. However, adding a trigger circuit to SCR device also increases the parasitic capacitance seen at the RFIN pad, which is difficult to handle for RF and MMW circuit applications [11].

2.2. Power-Rail ESD clamp circuit

Figure 4 shows the power-rail ESD clamp circuit which is the discharge path of current under ESD events. The circuit is composed of the RC-inverter-triggered NMOS, which is used to provide ESD current paths between VDD and VSS under ESD stress conditions. The R1 (~10k) and C1 (~10 pF) having a time constant of 0.1 s to 1 s can determine the ESD transients from the normal circuit operating conditions. The NMOS (MESD) with ~3000-m width is used as the main ESD clamp device. The large-sized NMOS (MESD) is turned on to provide ESD current path from VDD to VSS when positive ESD stress occurs from VDD to VSS. Likewise, when negative ESD stress from VDD to VSS, the parasitic diode in large- sized NMOS (MESD) also provides the ESD current path from VSS to VDD. The power-rail circuit does not contribute parasitic effects to RF input port since the power-rail ESD clamp circuit is placed between VDD and VSS [10].

2.3. Two-Stage low noise amplifier

Figure 5 shows the cascode configuration of LNA. This configuration provides a good input-output isolation and the transistor M2 isolates the Miller capacitance. Furthermore, the input impedance is obtained using the source degeneration inductor Ls, while the gate inductor Lg sets the resonant frequency [12]. To achieve simultaneous noise and power matching, the inductive source degeneration is used. The values of this inductor is designed for both gain and stability considerations. The main design parameters of the input stage are the sizes of M1 and M3, the source inductor, Ls and the gate inductor Lg of Figure 5. The sizes are then determined based on optimization to enhance impedance matching as well as achieve enough power gain [12].

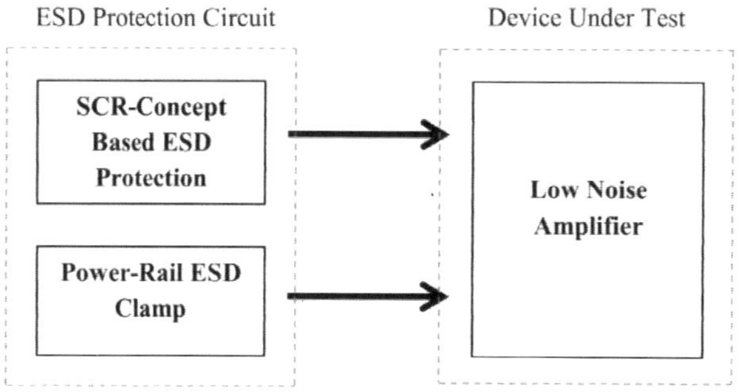

Figure 2. System Architecture

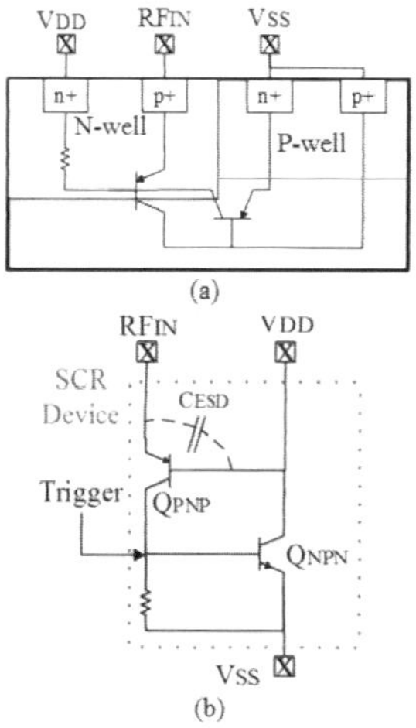

Figure 3. (a) Device Cross-sectional View (b) Equivalent Circuit of SCR

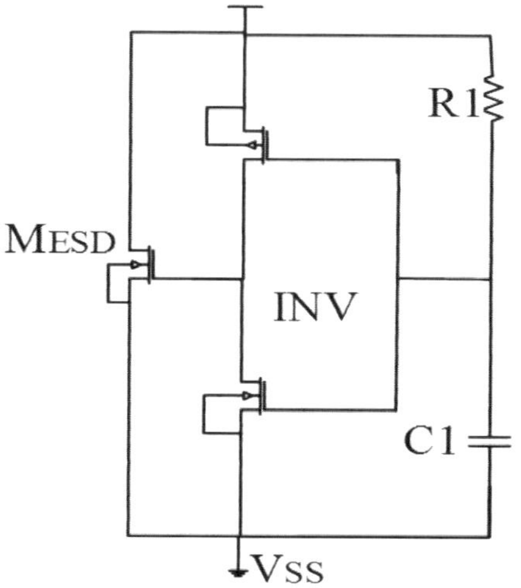

Figure 4. Power-Rail ESD Clamp Circuit

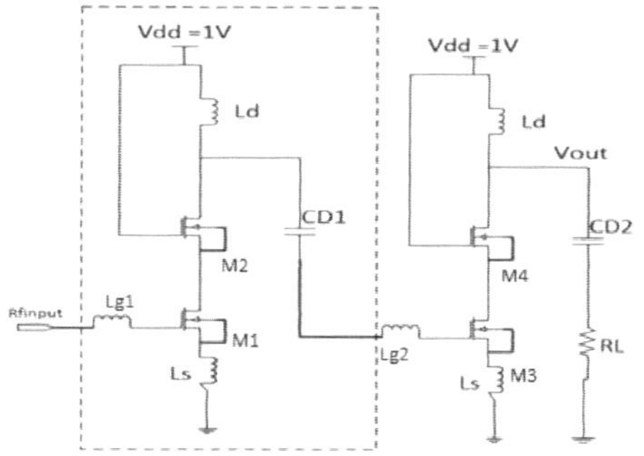

Figure 5. Power-Rail ESD Clamp Circuit

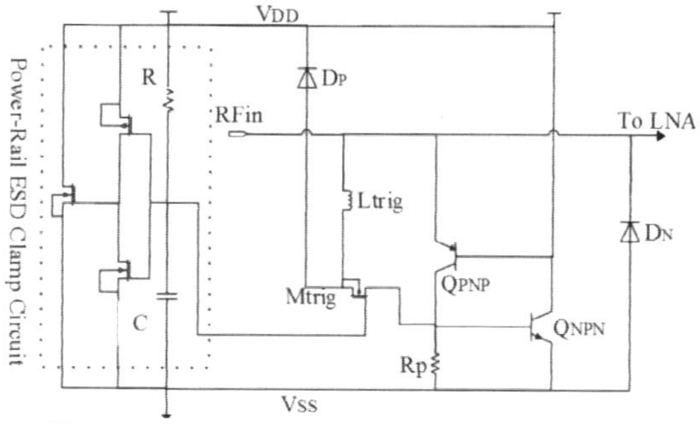

Figure 6. Proposed ESD Protection Circuit

2.4. Proposed ESD protection scheme

Figure 6 shows the proposed ESD protection circuit. The circuit uses an inductor-triggered SCR which consists of SCR architecture, an inductor, a MOS transistor and the RC-based ESD detection circuit coming from the power rail circuit block. The PMOS transistor exhibits the initial-on function for ESD protection, wherein, it can quickly pass the trigger signal to the SCR device. Under normal power-on conditions, the PMOS transistor is turned off to block the steady leakage current path from RFin to the trigger port of the SCR device. Meanwhile, during ESD stress conditions, the PMOS transistor is turned on so that the SCR device is quickly turned on to discharge the ESD current. The RC-based detection circuit coming from the power rail circuit is used to distinguish ESD stress conditions from normal circuit operating conditions with a time constant of 0.1us to 1us. The inductor is used to provide trigger path between RFin and trigger port of the SCR under ESD stress conditions. Moreover, it is used to compensate the parasitic capacitance of the SCR device (CESD). In addition, diodes Dp and Dn complete the proposed ESD protection circuit [10].

The dimensions of the inductor, PMOS transistor, SCR device and diodes Dn and Dp can be designed to minimize the RF performance degradation. Since the capacitor used in power-rail ESD clamp circuit is large enough (~10pF) to keep the node between R and C at ac ground, the impedance of trigger path seen at RFin to ground is calculated as;

$$Ztrig \approx j\omega Ltrig + 1/j\omega(Ctrig + Cdp) \quad (1)$$

$$Ztrig = j\omega(Ltrig - 1/(\omega^2(Ctrig + Cdp))) \quad (2)$$

where the ω is the angular frequency and the Ctrig can be expressed as

$$\text{Ctrig} \approx \text{Cgs} + \text{Cgb} + \text{Cdb} \quad (3)$$

,Cgs denotes gate-to-source capacitance, Cgb denotes gate-to- body capacitance and Cdb denotes drain-to-body capacitance of the PMOS transistor. The resonance angular frequency (ω_0), which is designed to be the operating frequency of RF signal, can be obtained by

$$\omega_0 = 1/((Ltrig - 1/(\omega^2(Ctrig + Cdp)))(Cesd)) \quad (4)$$

,where Cesd is the parasitic capacitance contributed by SCR and diode (Dn). Furthermore, solving for the value of Ltrig, the equation becomes

$$Ltrig = 1/(\omega^2 Cesd) + 1/(\omega^2(Ctrig + Cdp)) \quad (5)$$

The proposed circuit utilizes the impedance isolation tech- nique where it uses the inductor Ltrig to resonate out the parasitic capacitance generated by the ESD protection circuit. This technique is highly suitable for higher frequencies greater than 5-GHz and parasitic capacitance can be designed to be zero (ideally) at the desired frequency. The conventional ESD protection circuit consists of low capacitance diodes. The conventional diodes show that it is suitable only for lower frequencies (less than 5-GHz) but it is worst at high frequency applications. In contrast, the use of conventional diodes has smaller layout area compared to impedance isolation technique in the proposed circuit.

The proposed circuit is similar to the inductor-triggered SCR configuration but the only difference is the addition of diode Dp from the terminal of inductor from RFin pad to power supply VDD. The idea of adding a diode Dp comes from the circuit configuration where an ESD protection can be placed between one terminal of inductor to power supply VDD. From equation 5, the parasitic capacitance generated by diode Dp is inversely proportional to the size of the inductor Ltrig. Knowing that the inductor contributes the largest area in layout design, it is important that the size of the inductor is small to save layout area as well as cost of fabrication. Thus, with the addition of diode (Dp), it will decrease the size of the inductor resulting to smaller layout area. Furthermore, diode Dp will provide another discharging path of ESD current/voltage when ESD zapping (PD-Mode) occurs from RFin pad to power supply VDD.

Figure 7 shows the ESD current paths under positive- to-Vss (PS), positive-to-VDD (PD), negative-to-Vss (NS), and negative-to-VDD (ND) ESD stress conditions. During positive-to-Vss (PS) ESD stress, ESD current will first pass through the inductor and PMOS to trigger the SCR device. The major ESD current will be discharged by the SCR device from the RFin pad to Vss. Under positive-to-VDD ESD stress, the ESD current will be discharged by the parasitic diode path embedded in the SCR device from the RFin pad to VDD. During negative-to-Vss (PS) ESD stress, the ESD current will be discharged by the forward-biased Dn from the Vss to RFin pad. Under negative-to-VDD (ND) ESD stress, the ESD current will be discharged by the power-rail ESD clamp circuit and the Dn from VDD to RFin pad. The proposed ESD protection scheme can provide the corresponding current discharging paths with good ESD robustness.

IOP Conf. Series: Materials Science and Engineering **600** (2019) 012025 doi:10.1088/1757-899X/600/1/012025

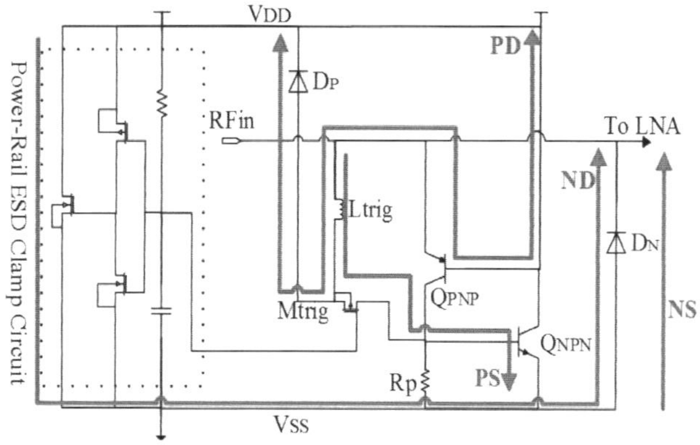

Figure 7. Proposed ESD Protection Circuit with ESD Zapping Modes

3. ESD simulations

3.1. S-Parameter simulation

The proposed ESD protection circuit will be simulated using the circuit in figure 8. The S-parameters S21 and S11 will determine the performance of the proposed ESD circuit design. The ideal value of S21 should be close to unity which indicates that no signal loss occurs during the simulation. Moreover, the desired value of S11 should be as small possible to have good impedance matching.

3.2. ESD robustness simulation

Figure 9 is used to simulate the HBM model. The Ch capacitor is estimated to be the capacitance that comes from the human body, while the resistor Rh is estimated to be the resistance of the hands of the body as the person touches the device. Each component will be thoroughly investigated if the threshold voltage generated by each device/component exceeds the breakdown voltages. The device under test is composed of the low noise amplifier with the proposed ESD protection circuit. For LNA, the transistor M1, M2, M3 and M4 will be examined as shown in fig. 5. Whereas, diodes Dp and Dn, Ptrig, PNP and NPN bipolar transistors, as well as the NMOS and PMOS transistor in inverter and MESD transistor of the power rail circuit in fig. 4. The device under test is simulated with different values of voltage source of 2-kV, 4- kV, 6-kV and 8-kV to test the robustness of the device.

4. Simulation results of proposed ESD circuit

One of the most challenging parts in this study is the designing of the ESD protection circuit. The researcher has tried implementing different techniques and schemes to cancel out the parasitic capacitance which degrades the whole perfor- mance of the device under test. In the end, the researcher used an inductor to resonate out the parasitic capacitance at a high frequency. The proposed circuit has been designed in 65-nm CMOS process.

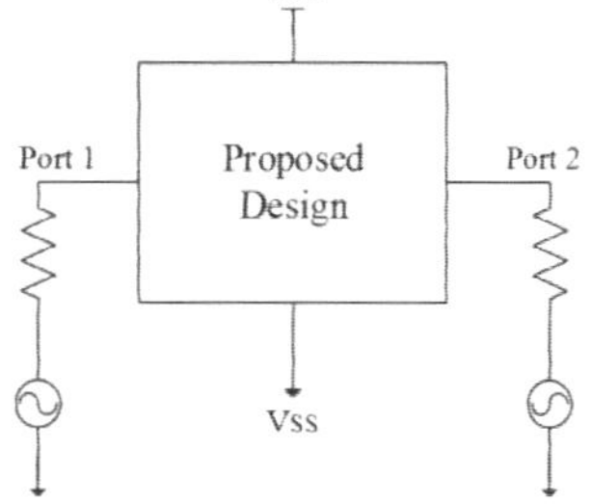

Figure 8. Schematic diagram for s-parameter simulation

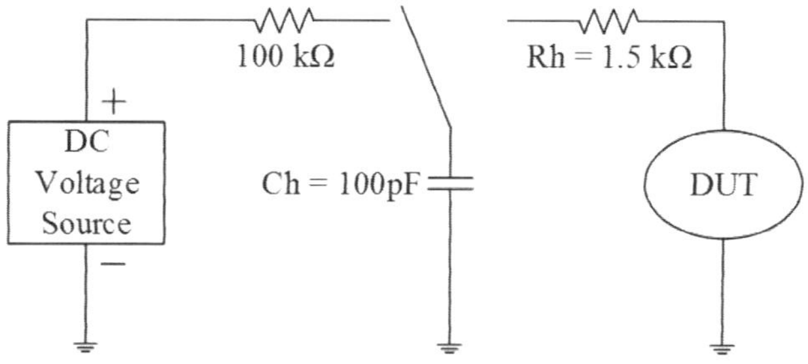

Figure 9. Schematic diagram for human body model simulation

4.1. Pre-Simulation

4.1.1 Circuit performance

Figure 10 shows the plot of the ESD protection circuits S parameters, obtained when the LNA circuit is not connected in the circuit. In the design, the ESD protection circuit achieved an S21 parameter of - 0.473mdB and S11 parameter of - 24.6dB. Due to the contribution of built-in parasitic capaci- tances of the circuit, the S21 parameter is not ideally equal to zero but it majorly cancels the parasitic capacitances. Furthermore, the ESD circuit is designed to operate at 60-GHz frequency. All the values of S11 parameter of the different process corners are less than -10 dB which indicates good impedance matching. Moreover, the S21 parameters have values that are close enough to each other which indicate a reliable performance under different process conditions.

4.1.2 ESD robustness

ESD induced failure is the effect when an ESD-sensitive device becomes part of the discharge path or when under the influence of high electrostatic field. One of the ESD induced failures is the dielectric breakdown. The researcher simulated the breakdown voltages of each component. According to Baker, the maximum electric field across a device gate is limited to 10MV/cm. This translates into 1V/10A of gate

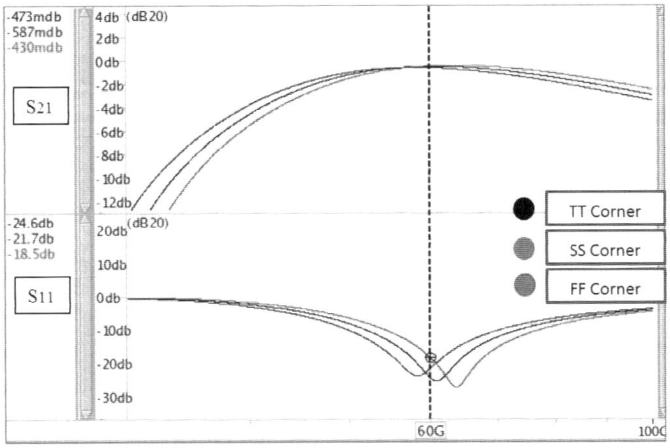

Figure 10. Graph of S-parameters (S21, S11)

oxide. Given in the BSIM 4 (v4.5) model of Synopsys Library, the CMOS transistor has tox of electrical gate equivalent oxide thickness of 30A [13]. Thus, under supply voltage of 1V, the maximum allowed DC voltage that can be applied across the gate oxide is 3V based on the calculation below.

$$(1V/10A) * 30A = 3V \tag{6}$$

Table 1. Gate oxide breakdown voltage of MOS transistors

	Low Noise Amplifier				ESD Protection Circuit			
Dielectric	M1	M2	M3	M4	MEsd	*Nch*	*Pch*	Ptrig
BV	$\|3V\|$	$\|3V\|$	$\|3V\|$	$\|3V\|$	$\|3V\|$	$\|3V\|$	$\|3V\|$	$\|3V\|$

Table 2. Reverse breakdown voltages of diodes

Reverse Breakdown Voltage of Diodes	
Dp	Dn
$\|10V\|$	$\|10V\|$

Table 3. Breakdown voltage of bipolar transistors

Breakdown Voltage of Bipolar Transistors	
NPN	PNP
$\|6.8kV\|$	$\|2.87kV\|$

Table 1 shows the gate oxide breakdown voltage of MOS transistors based on the calculation of equation 6. The threshold voltage *Vgs* of each transistor must not exceed the gate oxide breakdown voltage so that the transistor will not be damaged.

Table 2 shows the tabulated results of the reverse breakdown voltage of diodes Dp and Dn found in the proposed circuit. The reverse breakdown voltages of diodes are also shown in BSIM 4 (v4.5) Library used in simulating diodes. It is found out that, the reverse breakdown voltage is also -10V.

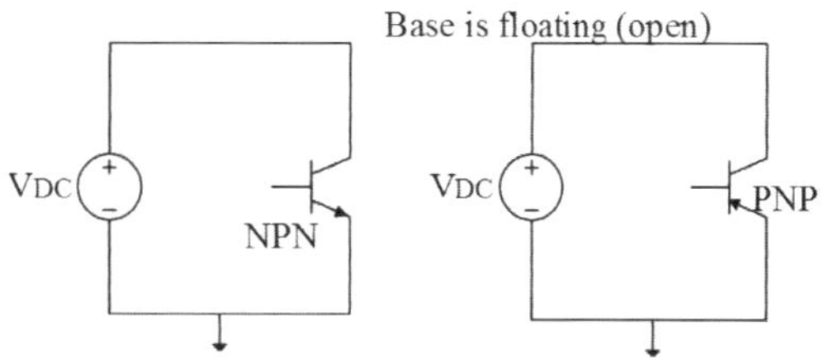

Figure 11. Circuit used in simulating the I-V characteristics curve of BJT

Table 4. Pre-simulation HBM ESD results

	Process Corner Variations		
	TT	SS	FF
PS HBM ESD Level (kV)	2	2	2
PD HBM ESD Level (kV)	6	4	8
NS HBM ESD Level (kV)	6	6	8
ND HBM ESD Level (kV)	4	4	4

The circuit used in simulating the I-V characteristic curve of BJTs is shown in figure 11. Based on simulations, the breakdown voltage BVCEO, between collector and emitter when base is open is tabulated in table 3.

Table 4 shows the pre-simulation results of the ESD robust- ness level of the device under different process corners. The results passed the 2-kV level military standard of robustness. These results come from the simulated circuit of figure 9.

A PS HBM ESD level of 2-kV means that the components of the proposed circuit withstand a 2-kV voltage supply without exceeding the breakdown voltages on each component. The breakdown voltages are determined for MOS transistor, diodes and BJTs of the proposed circuit with the LNA configuration. The same applies for PD Mode, NS Mode and ND Mode. Notice that all ESD zapping modes passed the 2kV military standard.

4.2. Post-simulation

4.2.1 Circuit Performance

Figure 12 shows the results of post-simulations of the proposed ESD circuit. Notice that, the TT, SS and FF process corners have values of S21 and S11 that are close to each other. Among the S21 parameters, the FF corner has the least value which is close to the desired value of unity. All of the S11 values are less than -10 dB which indicates that the circuit has good impedance matching.

4.2.2 ESD robustness

Table 5 shows the post-simulation results of Human Body Model. Notice that the results showed that at each process variation, it passed the military standard that HBM ESD level must be greater than or equal 2-kV. These results are simulated using figure 9. Like the pre-simulated values, the post-simulated values depend on the ability of each compo- nent/device (from the proposed ESD circuit with LNA) to withstand a large amount of voltage (2-kV, 4-kV, 6-kV and 8- kV). The component can withstand the voltage if the threshold voltage generated will not exceed to the breakdown voltages. For MOS transistors in LNA, the breakdown voltage is 3V, whereas for diodes in ESD protection circuit, the breakdown voltage is $|10V|$. For BJTs, the breakdown voltage is $|6.8kV|$and $|2.87kV|$ for NPN and PNP respectively.

Voltage generated will not exceed to the breakdown voltages. For MOS transistors in LNA, the breakdown voltage is 3V, for diodes in ESD protection circuit, the breakdown voltage is $|10V|$. For BJTs, the breakdown voltage is $|6.8kV|$and $|2.87kV|$ for NPN and PNP respectively.

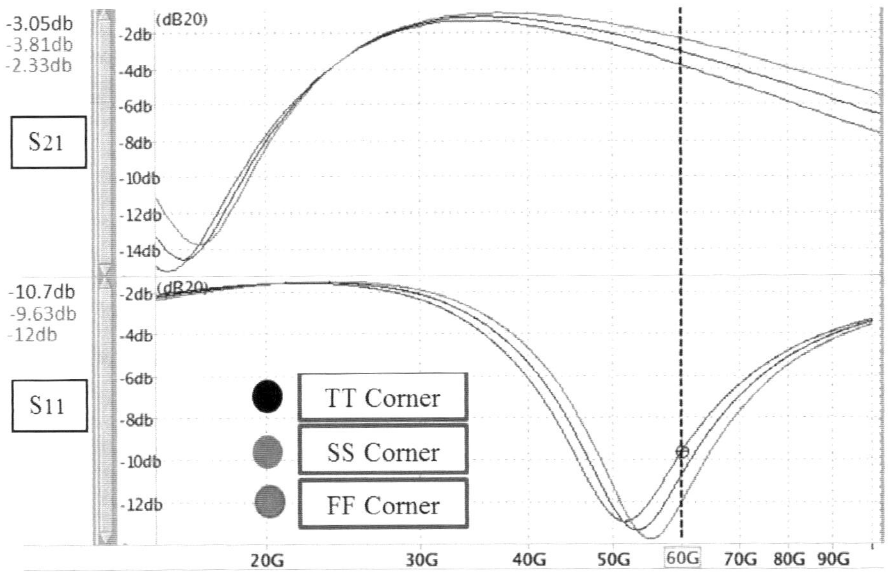

Figure 12. S-Parameters of Post-simulations of Proposed ESD Circuit

Table 5. Post-sim HBM ESD results

Process Corner Variations			
	TT	SS	FF
PS HBM ESD Level (kV)	2	2	4
PD HBM ESD Level (kV)	2	4	4
NS HBM ESD Level (kV)	2	8	8
ND HBM ESD Level (kV)	2	6	8

Table 6. Presim versus Post-sim HBM ESD results

Process Corner Variations						
	Pre-simulation			Post-simulation		
	TT	SS	FF	TT	SS	FF
PS HBM ESD Level (kV)	2	2	2	2	2	4
PD HBM ESD Level (kV)	6	4	8	2	4	4
NS HBM ESD Level (kV)	6	6	8	2	8	8
ND HBM ESD Level (kV)	4	4	4	2	6	8

4.3. Pre-simulation versus Post-simulation

Figure 13 shows the pre-simulation and post-simulation of the S-Parameters of the ESD protection circuit. For S21 parameter, the pre-simulated value is -472mdB which is close to the ideal value of unity or 0-dB. A 0-dB value of S21 parameter means that there is no loss in applying the ESD Protection circuit. The post-simulation value is -3.1dB which is mainly due to the parasitic capacitance and parasitic inductance generated by the layout design of the circuit. Furthermore, the S11 parameter must be as small as possible to have good impedance matching. Since both the pre-simulation and post-simulation values of S11 are less than -10dB, then, both results indicate good impedance matching.

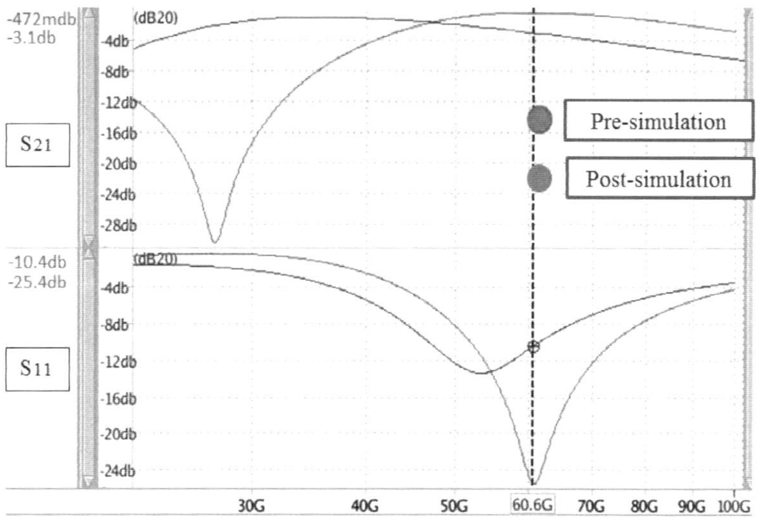

Figure 13. Pre-simulation and Post-simulation Results of ESD Protection Circuit

Table 6 shows the pre-simulation and post-simulation values of HBM level. From the results, all corners pass the 2-kV requirement of the military standard. It can be inferred that the proposed ESD protection circuit can withstand 2-kV of power supply if fabricated.Figure 14 shows the pre-simulation and post-simulation of the LNA performance with ESD circuit. The pre-simulated value of S21 parameter is 13.4dB. However, the post-simulated value becomes 12.4dB. The results showed that there is a loss in gain or S21

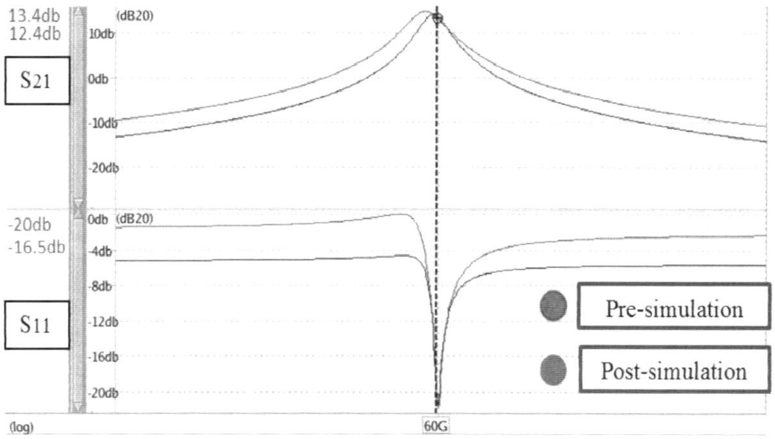

Figure 14. Pre-simulation and Post-simulation Results of LNA with ESD Circuit

parameter by 1dB. This is mainly due to the parasitic capacitances introduced by the ESD protection circuit. Moreover, the post-simulation value of S11 parameter of the LNA with ESD circuit becomes higher from -20dB to-16.5dB. Same reason applies as the S21 parameter which is caused mainly by the parasitic capacitance introduced by the ESD protection circuit. Parasitic capacitances from the ESD protection circuit will cause signal loss of the RF input signal.

4.4. Assessment between Ltrig, Dp and Dn

To show the relationship of the components of the proposed circuit and its impact to its performance, an assessment is needed to verify the concepts presented in related literature and methodology. Table 7 shows three cases where the val- ues of Ltrig, Dp and Dn are changed to its optimum ESD performance level which is designed at 60-GHz frequency. Notice that as the value of Ltrig increases, the value of Dp and Dn decreases as shown in the decrease in value of multiplier. Thus, a larger value of Ltrig is needed to cancel out the parasitics generated by smaller Dp and Dn. However, as the device/component becomes larger in size, it increases the parasitic capacitance but it becomes more robust to ESD stress/events [10]. Figure 15 shows the simulated S21 and S11 parameters of the three different cases presented. As observed in figure 15, all cases are designed to efficiently function at 60-GHz frequency. Table 8 summarizes the results shown in figure 15.Table 8 shows the tabulated results of figure 15. The ideal value of S21 is unity or 0dB, whereas S11 should be small enough to have good impedance matching. Among all cases, case 3 offers the best performance having the closest S21 parameter of the ESD protection circuit to 0dB. Also, case 3 has the smallest S11 parameter. However, the drawback of case 3 is the increasing size of the inductor which poses the highest layout area of approximately 150um x 180um. In contrast, the ESD circuit used in this paper is case 1 which has the least S21 parameter but comparable to the other cases. Also, all three cases pass the requirement of S11 parameter that is -10 dB, to have a good impedance matching according to [10]. Thus, from the results, the performance of the ESD protection circuit is inversely proportional to the layout area knowing the inductor has the largest exposure of area. As the value of inductor increases, the performance of ESD circuit is better but the layout area increases.

Table 7. Different values of Ltrig, DP and DN

	Ltrig (nH)	Dp	Dn
Case 1	0.11	W/L = 0.9u/20u, M = 9	W/L = 0.9u/20u, M = 3
Case 2	0.2	W/L = 0.9u/20u, M = 4	W/L = 0.9u/20u, M = 2
Case 3	0.2	W/L = 0.9u/20u, M = 1	W/L = 0.9u/20u, M = 1

Table 8. Simulation Results on S21, S11 and area

	S21	S11	Layout Area of ESD
Case 1	-473 mdB	-24.6 dB	95um x 120um
Case 2	-339 mdB	-27.9 dB	125um x 150um
Case 3	-257 mdB	-30.7 dB	150um x 180um

4.5. Layout Design

Figure 16 shows the layout design of the ESD protection circuit without the power rail circuit. The layout area of the proposed design is 95um x 120um. Figure 17 shows the whole layout of ESD protection circuit. It can be observed from the figure that the area mainly depends on the size of the inductor and the capacitor.

4.6. Comparison among ESD protection circuits

Table 9 shows the comparison results of the designed ESD protection circuit compared with other references. Among all ESD protection circuits presented, Ref. [10] has the best S21 parameter while Ref. [8] has the best S11 parameter. However, the layout of the proposed circuit has 12% reduction in area of Ref. [8] and Ref. [10] papers. Among the ESD protection circuits, this study has the lowest area but also has the least S21 parameter of -3.1 dB. Furthermore, among all ESD circuit, this work has the highest value of S11 parameter with the value of -10.4 dB. Nevertheless, it still has good impedance matching. The performance of the proposed ESD protection circuit is dependent on the parasitic capacitance and parasitic inductance of the different elements in the circuit itself. Moreover, the proposed ESD protection circuit passed the military standard requirement of ESD robustness which is 2-kV based on simulations.

IOP Conf. Series: Materials Science and Engineering **600** (2019) 012025 doi:10.1088/1757-899X/600/1/012025

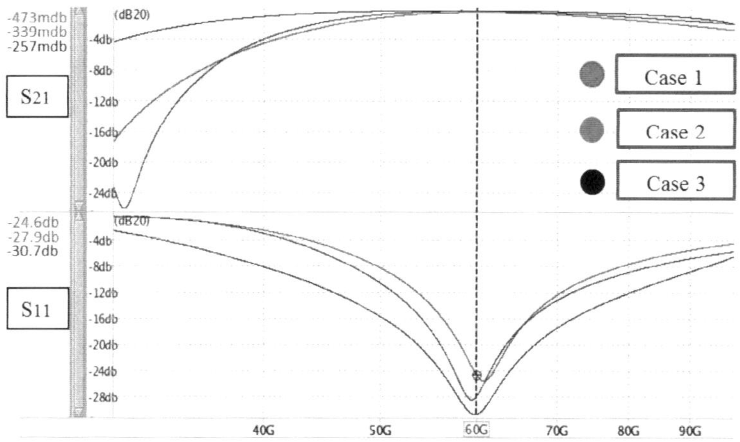

Figure 15. Simulated S21 and S11 Parameters of Three Cases

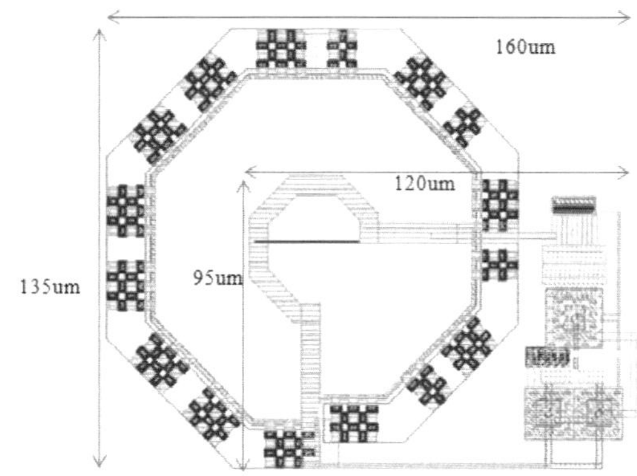

Figure 16. Layout Design of ESD Protection Circuit without Power Rail Circuit

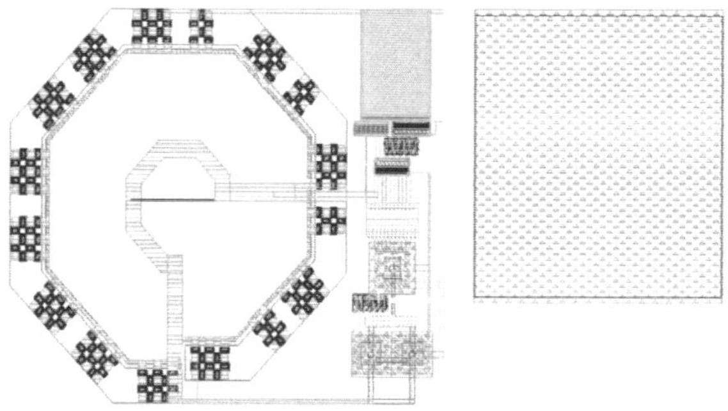

Figure 17. Layout Design of ESD Protection Circuit with Power Rail Circuit

5. Conclusion

This study completes the design of a compact ESD pro- tection circuit for V-band RF applications in a 65-nm CMOS process. The ESD protection circuit is developed to help RF circuit designers for them to easily apply ESD protection circuit in 60-GHz RF circuits. The proposed ESD protection circuit uses an inductor triggered SCR concept based circuit as well as ESD protection diodes which are provided in the commercial CMOS process. The ESD protection circuit is then applied to the low noise amplifier circuit to determine its performance.However, for an integrated circuit to be reliable; the IC should be capable to manage the ESD events that occur from manufacturing process to end user. Testing and characterizing the circuits to guarantee the reliability of ESD circuits is one concern for designers. In the study, the robustness of ESD protection circuit was simulated and can achieve the 2-kV HBM ESD level. Based on the simulated result, the ESD protection circuit can achieve the required 2-kV ESD robustness condition in military standard if fabricated. In addition, the proposed ESD protection circuit is compact and area efficient, which is a useful solution for on chip ESD protection design in 60-GHz RF applications to save fabrication cost. The layout chip area of the ESD protection circuit is 95um x 120um which has 12% reduction of area compared to related references. Despite of having an efficient area, the ESD protection circuit attained a low S11 parameter of less than -10 dB which indicates good impedance matching but due to parasitic capacitance and parasitic inductance in the layout process, the simulated S21 parameter is -3dB which is presents a drawback in the study. Nevertheless, this result is comparable due to the LNA with ESD performance which has small degradation in performance. Therefore, the proposed ESD protection scheme can be widely used to achieve high ESD robustness within small layout simultaneously.

Acknowledgment
The researchers would like to thank the Microelectronics Laboratory of MSU-IIT and the Department of Science and Technology - Philippine Council for Industry, Energy, and Emerging Technology Research and Development (DOST-PCIEERD) for the integrated circuits design tools being provided for this research.

References

[1] D. Raju, *Study of ESD Effects on RF Power Amplifiers*, May 2011.
[2] R. Strube, *Challenges of Electro-Static Discharge (ESD) in the Electron- ics Industry*, Retrieved from http://www.uic.com/challenges-of-electro- static-discharge-esd-in-the-electronics-industry/ 2015.
[3] A. Wang, *On-Chip ESD Protection for Integrated Circuits, an IC Design Perspective*, Norwell, MA: Kluwer Academic Publishers, 2002.
[4] A. Wang, *ESD + RFIC Co-Design for Whole-Chip DesignOptimization*, Retrieved from http://www.asqed.com/English/Archives/2010/Tutorials/ 2010.
[5] M. Ker, C. Lin, Y. Hsiao, *Overview on ESD protection designs of low parasitic capacitance for RF ICs in CMOS technologies*, IEEE Trans. Device Mater. Reliab., vol. 11, no. 2, pp. 207218, 2011.
[6] W. Soldner, M. Streibl, U. Hodel, M. Tiebout, H. Gossner, D. SchmittLandsiedel, J. Chun, C. Ito, R. Dutton, *RF ESD protection strategiesCodesign vs. low-C protection*, Microelectron. Reliab., vol. 47, no. 7, pp. 10081015, 2007.
[7] V. Vassilev, S. Thijs, P. Segura, P. Leroux, P. Wambacq, G. Groe- seneken, M. Natarajan, M. Steyaert, H. Maes, *Co-design methodology to provide high ESD protection levels in the advanced RF circuits*, in Proc. E OS/ESD Symp., pp. 19, 2003.
[8] C. Lin, L. Chu, M. Ker, *ESD protection design for 60-GHz LNA with inductor-triggered SCR in 65-nm CMOS process*, IEEE Trans. Microw. Theory Tech., vol. 60, no. 3, pp. 714723, 2012.
[9] M. Ker, K. Hsu, *Overview of on-chip electrostatic discharge protection design with SCR-based devices in CMOS integrated circuits*, IEEE Trans. Device Mater. Rel., vol. 5, no. 2, pp. 235249, 2005.
[10] C. Lin, L. Chu, S. Tsai, M. Ker, *Design of Compact ESD Protection Circuit for V-Band RF Applications in a 65-nm CMOS Technology*, 2012.
[11] L. Chu, C. Lin, M. Ker, *Design of Dual-Band ESD Protection for 24-/60-GHz Millimeter-Wave Circuits*, 2013.
[12] A. Karanicolas, *A 2.7-V 900-MHz CMOS LNA and Mixer*, IEEE J. Solid-State Circuits, vol. 31, pp 1939 1944, 1996.
[13] R. Baker, *CMOS: Circuit Design, Layout, and Simulation*, Third Edition, 2010.
[14] C. Lin, L. Chu, M. Ker, *Design and implementation of configurable ESD protection cell for 60-GHz RF circuits in a 65-nm CMOS process*, Microelectron. Reliab., vol. 51, n o. 8, pp. 13151324, Aug. 2011.

IOP Conf. Series: Materials Science and Engineering **600** (2019) 012026 doi:10.1088/1757-899X/600/1/012026

Synchronous condenser excitation system parameter identification based on improved artificial bee colony algorithm

Nan Li[1], Lian Meng[1] and Bin Geng[2]

[1]Department of Electric Engineering, School of Northeast Electric Power University, Jilin City, Jilin Province, 132012, China
[2]State Grid Jilin Electric Power Co., Ltd. Changchun City, Jilin Province, 132021, China

jllinan@163.com

Abstract. The quality of excitation control of the synchronous condenser directly determines the reactive power regulation ability and control support performance of the power grid.Aiming at parameter identification of non-linear links such as limit and saturation in the synchronous condenser excitation system, a parameter identification method of the synchronous condenser excitation system based on improved artificial bee colony algorithm is proposed.Firstly, the order of the phase modulation operation is deduced, and the mathematical model of the synchronous condenser excitation system is established based on it.Then the artificial bee colony algorithm is improved and applied to the parameter identification of the synchronous condenser excitation system. A parameter identification strategy based on sensitivity analysis is proposed for the unstable and multi-solution situation of multi-parameter identification results. Analysis of parameter sensitivity and identification based on parameter sensitivity. Finally, the identification parameters and the overall identification results obtained after sensitivity analysis are compared and analyzed. The results show that the proposed method can effectively identify the parameters of the non-linear link and has strong robustness. It can provide accurate models and parameters for power system stability analysis.

1. Introduction

With the extensive application of UHVDC transmission, especially the voltage stability problem.It is mainly manifested in the insufficient dynamic reactive power support during the fault period, which objectively requires large-scale DC active power transmission and must match large-scale dynamic reactive power[1]. At present, synchronous condenser, static var compensator (SVC) and static synchronous compensator (SVG/STATCOM) are mainly used in large-scale reactive power compensation devices in power system[2,3].To study the influence of the access of the synchronous condenser on the stability of the power system, it is necessary to have a suitable synchronous condenser excitation system model and accurate parameters .Therefore, it is of great theoretical and practical value to study an effective method for identifying model parameters of the synchronous condenser excitation system.

The synchronous condenser is essentially a synchronous motor with excitation control and no mechanical load. All its electrical performance is controlled by excitation system.The excitation control parameters of the synchronous condenser directly determine the reactive power regulation and control support performance of the synchronous condenser[4,5].Literature[6] By systematically analyzing the influence of various electrical parameters on the transient and sub-transient characteristics of the synchronous condenser, the key technical parameters and measures for optimizing the dynamic performance of the synchronous condenser are extracted.In reference [7], the model and parameters of the field excitation system are introduced into the stability calculation of power grid, which improves the authenticity and precision of the simulation research.At present, there are some studies on the electrical parameters of the phase modulator, but there are relatively few studies on the parameter identification of the synchronous condenser excitation system.With the development of identification technology, great progress has been made in the model and parameter identification of generator excitation system. These identification methods can be used as reference for parameter identification of synchronous condenser excitation system.Literature [8,9] identifies the non-linear link of generator excitation system using genetic algorithm, which proves the feasibility of using artificial intelligence algorithm to identify excitation system parameters.In reference [10], adaptive particle swarm optimization (APSO) is applied to identify the parameters of excitation system, which improves the convergence rate.These artificial intelligence algorithms have some advantages in solving complex non-linear parameter identification, but there are still some problems, such as slow convergence speed, easy to fall into local optimum, unstable identification results and so on.

In view of the above problems, this paper combines the mathematical model of generator excitation system. Firstly, according to the operation mode, derivation of transfer function order in phase modulation operation. On the basis of constructing the mathematical model of the synchronous condenser excitation system, an improved artificial bee colony algorithm is proposed to identify the parameters of the non-linear link of the model.Then, by analyzing the trajectory sensitivity of each parameter under disturbance, a parameter identification strategy based on sensitivity analysis is proposed.Finally, simulation experiments are used to verify the effectiveness and accuracy of the proposed method.

2. Construction the model of the synchronous condenser excitation system

Although the excitation mode of the synchronous condenser is the same as that of the generator, the structure and electrical characteristics of the synchronous condenser and the generator are different.This paper refers to Professor Liu X,L's linearized model based on the fifth-order description of synchronous generators, taking into account the effect of damper winding[11].As shown in Figure 1.

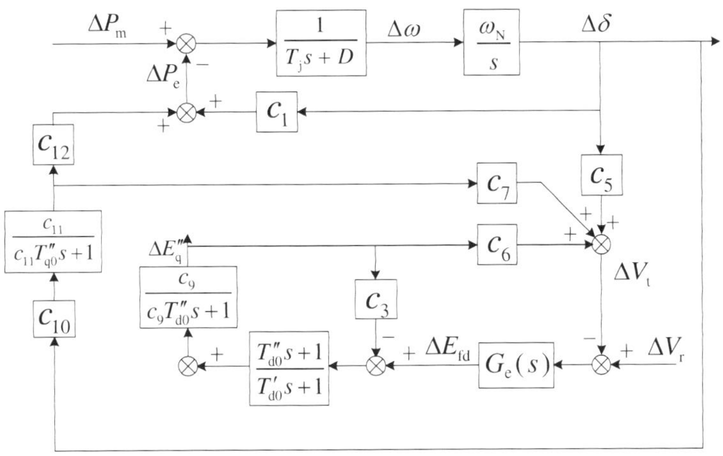

Figure 1. Linearized C1~C12 model for single machine infinite system.

Because there is no no idle running state of the synchronous condenser, the active angle and internal power angle are equal to zero. According to this condition, the above model is simplified, and the simplified model diagram of the synchronous condenser is obtained as shown in Figure 2.

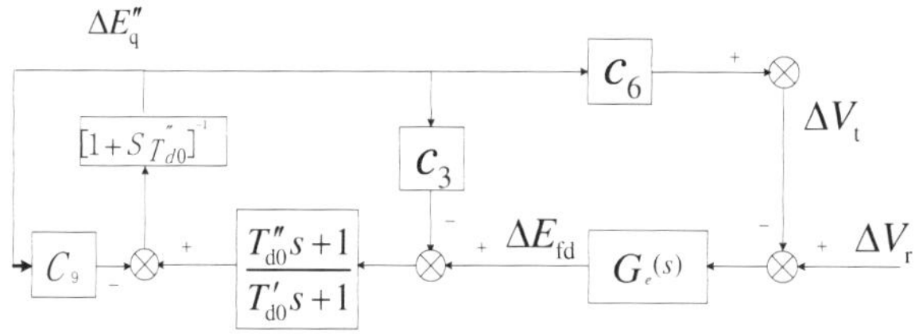

Figure 2. Simplified synchronous condenser phased operation model block diagram.

In order to obtain the relationship between the terminal voltage and the excitation potential and determine the order of the phase modulation operation of the phase modulator, the following deductions are made:

$$\frac{\Delta V_t}{\Delta E_f}=\frac{c_9(1+T''_{d0}S)*c_6}{(1+T'_{d0}S)(1+c_9T''_{d0}S)+c_3c_9(1+T''_{d0}S)} \tag{1}$$

To simplify the formula:

$$\frac{\Delta V_t}{\Delta E_f}=\frac{\dfrac{c_9c_6}{1+c_3c_9}}{1+\dfrac{T'_{d0}S}{1+c_3c_9}} \tag{2}$$

Order

$$\mathrm{k1}=\frac{c_9c_6}{1+c_3c_9};\ \mathrm{k2}=\frac{1}{1+c_3c_9} \tag{3}$$

$$\frac{\Delta V_t}{\Delta E_f}=\frac{k1}{1+k2*T'_{d0}S} \tag{4}$$

According to the above deduction, the first-order function of the phase modulation operation of the phase modulator can be obtained. Based on this, the transfer function block diagram of the excitation system of the phase modulator can be established in MATLAB as shown in Figure 3.

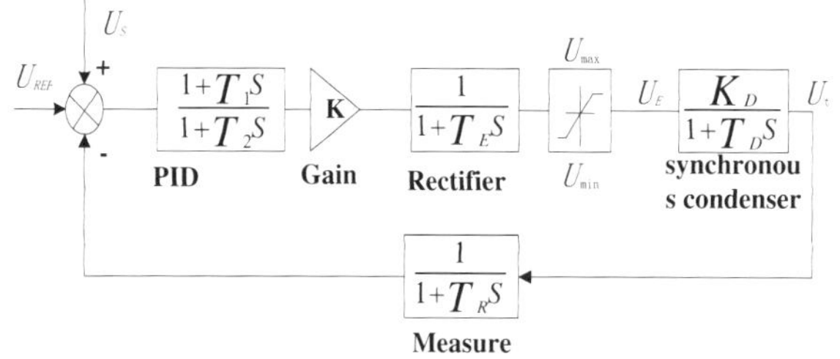

Figure 3. Transfer function block diagram of synchronous condenser excitation system.

In Fig 3, U_{Ref} is the input disturbance, U_E is the excitation terminal voltage and U_T is the phase modulator terminal voltage. In general, the design values are used for the parameters in the model. The parameters that need to be calculated or identified are PID link time constants T_1 and T_2, gain K of integrated amplification link, excitation time constant T_E, and phase modulator parameters K_D and T_D.

IOP Conf. Series: Materials Science and Engineering **600** (2019) 012026 doi:10.1088/1757-899X/600/1/012026

3. Parameter identification of synchronous condenser excitation system based on IABC

3.1. Improved artificial bee colony algorithm

The artificial bee colony algorithm(ABC)has a strong advantage in dealing with multi-parameter optimization process [13].In order to speed up the convergence speed of the algorithm and reduce the probability of the algorithm jumping out of the local extremum, the search neighborhood of the new honey source is improved in the algorithm, and an intermediate factor is added. If the adaptation of the previous generation is worth updating, it shows that the search direction tends to be correct, then the search direction is continued; on the contrary, the search direction tends to be opposite.The specific operation is shown in the formulas (5, 6).

When the honey source of the previous generation is updated, let n=1, otherwise n=0; m record the positive and negative sign of the previous generation φ_b .

$$z_{i,j} = x_{i,j} + \varphi_a (x_{best,j} - x_{i,j}) + \varphi_b |(x_{c1,j} - x_{i,j})| \tag{5}$$

$$\varphi_b = \begin{cases} m * rand(0,1) & ,n = 1 \\ -m * rand(0,1) & ,n = 0 \end{cases} \tag{6}$$

In the formula, $z_{i,j}$ is the renewal solution; $X_{i,j}$, $X_{c1,j}$ is the random solution corresponding to two different food sources; $X_{best,j}$ records the current optimal solution of individuals in the population; φ_a , φ_b is the random number between [-1,1].

The improved ABC algorithm introduces a random perturbation component into the search formula (5), which can randomly change the search area and jump out of the local optimum.Then, the intermediate factor is added to the formula(6). By comparing with the previous generation honey source, the algorithm consciously develops the region rich in honey source and speeds up the convergence speed of the algorithm.

3.2. Parameter identification strategy based on sensitivity analysis

As a complex rotating element, the synchronous condenser belongs to a high-order system in parameter identification.The sensitivity of different parameters under the same disturbance must be different, and different identification order will lead to different identification results[15].Therefore, this paper considers classifying the parameters according to the sensitivity level, fixing the parameters with small sensitivity as the design value, focusing on identifying the parameters with high sensitivity. For the overall transfer function of the synchronous condenser model, the sensitivity of the parameters of the transfer function is defined as:

$$H(\theta, s) = \lim_{\theta \to 0} \frac{\Phi(\theta + \Delta\theta, s) - \Phi(\theta, s)}{\Delta\theta} \tag{7}$$

In the formula, Φ is the transfer function of the system and $\Delta\theta$ is the variation of the parameters.

The trajectory sensitivity (relative value) formula of sensitivity parameter θ_j of power system is as follows:

$$\frac{\partial[y(\theta,k)/y_0]}{\partial[\theta_j/\theta_{j0}]} = \frac{[y_1(\theta_1,...,\theta_j + \Delta\theta_j,...,\theta_m,k)}{2\Delta\theta_j/\theta_{j0}} - \frac{y_2(\theta_1,...,\theta_j - \Delta\theta_j,...,\theta_m,k)]/y_0}{2\Delta\theta_j/\theta_{j0}} \tag{8}$$

In the formula, θ_{j0} is the given value of the parameter θ_j and y0 is the corresponding steady-state value θ_{j0} .

4. Analysis of simulation examples

The synchronous condenser excitation system as shown in Fig. 4 is built in Matlab, and the theoretical values of the parameters to be identified are set according to the range of parameters provided by the manufacturer.theoretical value:K=200, T1=1, T2=4, TE=0.01, KD=0.29, TD=3.58, Upper and lower limit of limit link:UAMAX=6, UAMIN=-2.

The synchronous condenser excitation system include linear and non-linear parameters. When the excitation system is in a small disturbance (generally less than 5%), the system is in a linear stage. The model can be identified directly.However, when the disturbance is large (generally more than 20%), the output voltage will be limited to a certain extent (in a non-linear state), as shown in Figure 4, the limiting link will play a role in this state.

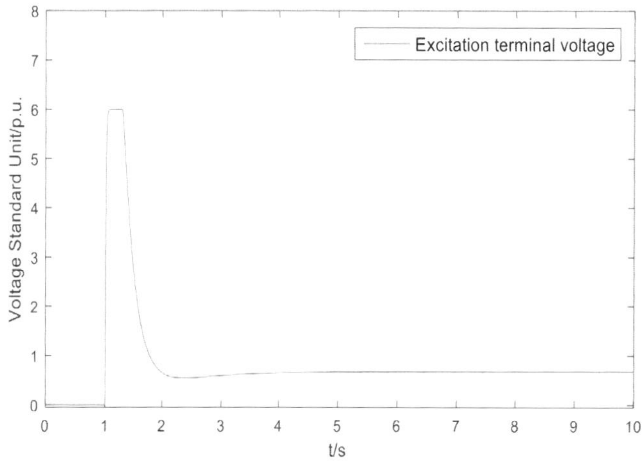

Figure 4. Exciter voltage output map.

Firstly, a 5% step disturbance signal is applied to the signal synthesis port 1s to identify the parameters of the linear link.Then,the identified parameters K, T_1, T_2 and T_E of the linear link are taken as known.The input of the system is a large step disturbance with a unitary value of $\pm$ 0.2. The output of the synchronous condenser is the output of the system. Identify U_{AMAX} and U_{AMIN}.

In order to verify the superiority of the improved bee colony algorithm in parameter identification, the improved artificial bee colony algorithm and other intelligent algorithms ((GA), (PSO), (DE), (ABC) are used to identify the parameters as a whole.The search interval of each parameter is within the range of [-20%, +20%].The parameters of the algorithm are as follows: population size Np = 80, population dimension N = 6, maximum iteration number GM = 50.The simulation output is shown in Figure 5., the identification results of each algorithm are shown in Table 1, and the convergence curve is shown in Figure 6.

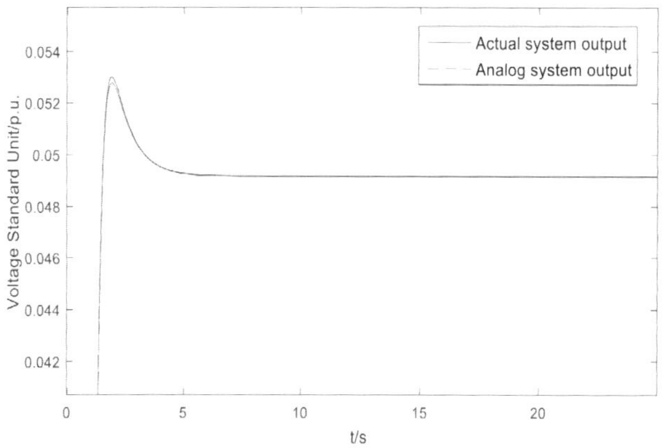

Figure 5. Synchronous condenser output voltage curve.

Figure 5. shows that the identification curve of the improved ABC algorithm fits well with the theoretical output curve. In order to better compare the algorithms, the average error of the identification results of each algorithm is analyzed as shown in Table 1.

Table 1. Overall identification results under different algorithms.

Parameter	Mean deviation	K	T_1	T_2	T_E	K_D	T_D	U_{AMAX}	U_{AMIN}
Design value		200	1	4	0.01	0.29	3.58	6.0	-2.0
GA	9.47%	189.95	0.901	4.183	0.0083	0.317	3.181	5.980	-2.013
PSO	9.41%	189.44	1.093	3.882	0.0115	0.302	4.222	6.010	-1.991
DE	9.39%	189.25	1.090	3.890	0.012	0.311	4.022	6.012	-2.002
ABC	9.16%	190.74	1.195	4.070	0.012	0.278	3.761	6.005	-1.995
IABC	8.22%	211.19	1.030	4.481	0.012	0.289	3.278	6.003	-1.994

By comparing the average deviation of the identification results (the average of the total relative errors between the parameter identification value and the design value), it can be seen that the average error of the improved ABC algorithm is smaller than that of other algorithms, and the identification accuracy is relatively high.

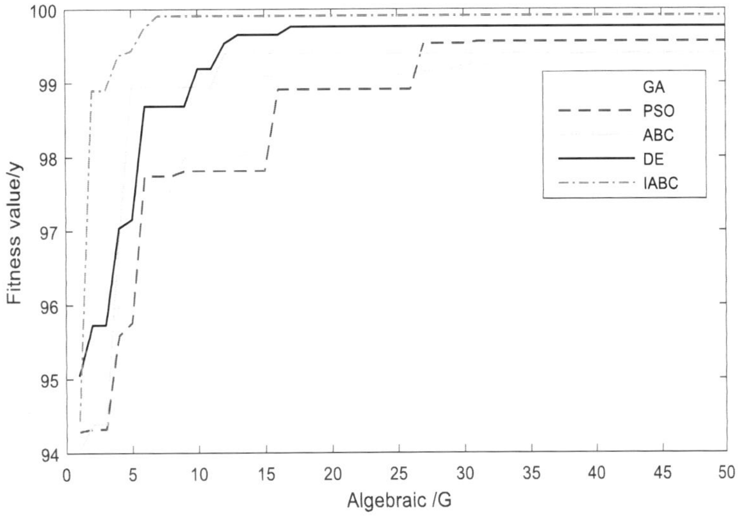

Figure 6. Convergence characteristic curve of algorithms.

From the figure 6, it is more intuitive to see that GA and PSO algorithms are easy to fall into local optimum, and the convergence speed is slow, and the effect of dealing with multi-dimensional complex parameter identification problem is poor.The global search ability of DE and ABC algorithm is strong, but there are also premature problems, which need further improvement.Compared with ABC algorithm, the improved ABC algorithm has better convergence characteristics, can jump out of local optimum faster, and has obvious advantages in multi-parameter identification.

Although there is little difference between the identified trajectory curve and the theoretical value, it can be found that the relative errors of each parameter are quite different, and some of them even exceed the allowable range of errors by analyzing the relative errors of each parameter.At the same

time, it is found that the results of multi-parameter global identification are not very different, but the identification results are unstable. It can be seen that the identification results of the whole system are unstable and the errors of individual parameters are large.For this reason, an isolation identification strategy based on sensitivity analysis is proposed to identify the model parameters.
The trajectory sensitivity of the synchronous condenser parameters is calculated.Then calculate the trajectory sensitivity of each parameter. The result is shown in the figure 7.

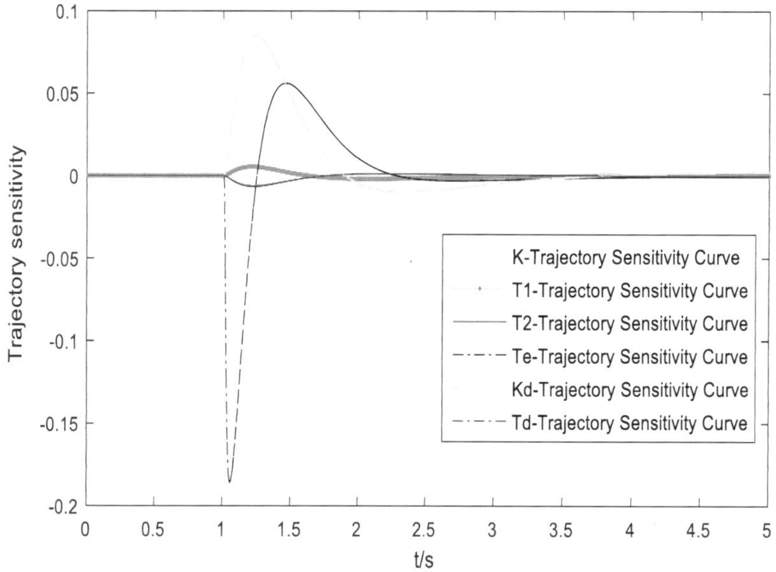

Figure 7. Parameters trajectory sensitivity curve.

Table 2. Average sensitivity of each parameter.

Parameter	K	T_1	T_2	T_E	K_D	T_D
Average sensitivity	0.0656	0.0301	0.0435	0.0054	0.0656	0.0421

It can be seen from the table that the sensitivity of parameters K and K_D is equal, while that of T_2 and T_D is close.Therefore, there may be some implicit function relationship between the two sets of parameters,which can be identified separately.The sensitivity of T_E is small, indicating that it is insensitive and can be used as a steady-state parameter.According to the structure of the model, the parameters K_D and T_D which are close to the measuring point are selected for identification.Then the values of parameters K_D and T_D are taken as known parameters. Identificate the parameters of K、T_1 and T_2 .Finally, the results of sensitivity analysis are shown in Table 4. Its objective function value is 99.989.

Table 3. Identification based on sensitivity analysis.

Parameter	K	T_1	T_2	T_E	K_D	T_D
Design value	200	1	4	0.01	0.29	3.58
Identification value	199.39	0.974	3.914	0.001	0.289	3.561
Relative error	0.31%	0.26%	2.15%	0	0.34%	5.30%

The traditional global identification results are as follows: Table 4, It is compared with the results identified by sensitivity analysis as shown in Table 4 below.

Table 4. Comparison of different identification strategies.

Identification strategy	Mean deviation	K	T_1	T_2	T_E	K_D	T_D
Global Identification results	8.22%	211.19	1.030	4.481	0.012	0.0289	3.278
Distribution identification results	1.39%	199.39	0.974	3.914	0.001	0.289	3.561

From the comparison of the above simulation identification results, it can be found that there is a big gap between the overall identification results and the identification results based on sensitivity analysis. From the point of view of the objective function, after sensitivity analysis, its accuracy is higher, and its results are more in line with the requirements of the operation state.

References

[1] Pourbeik P, Koessler R J, Quaintance W and Wong W 2006 *Conf. on Power Engineering Society General* (Montreal)
[2] Mithulananthan N, Canizares C, Reeve J and Rogers G 2003 *IEEE Trans. Power Syst.* **18** 786-92
[3] Teleke S, Abdulahovic T, Thiringer T, and Svensson J 2008 *IEEE Trans. Power Del.* **23** 1606-12
[4] Zhang K, Cui Y, Zhuang W, Miu Y, Yang Z, Feng Y, Zhang Q and Yu Y 2017 *Power Sys Prote Cont.* **22** 144-8
[5] Wang Y, Zhang Y, Zhou Q, Li Z, Jiang Y and Tang Y 2017 *Power Sys Techno* **01** 28-34
[6] Li Z, Jiang W, Wang Y, Li W, Li C and Xian Z 2017 *Large Motor Technology* **4** 15-22
[7] Pan W, Liu W, Li Y, Yang Y, Shi K and Fan Yue 2008 *Power Sys Techno* **11** 34-39
[8] Lee K and Yang F 1998 *IEEE Trans. Power Syst.* **1** 108-111
[9] Jiang P, Wang X, Wang Y, Xu K and Huang S 2008 *Power Automation Equipment* **6** 31-35
[10] Wildenhues S, Rueda J L and Erlich I 2015 *IEEE Trans. Power Syst.* **5** 2538-2546
[11] Liu X L, Liu Z and Gong H G 2000 *Proc Chin Soc Electrical Eng* **10** 20-23
[12] Ding Z H, Lu Z R and Liu J K 2018 *Science China Technological Sciences* **3** 1-10
[13] Sapkota B and Vittal V 2010 *IEEE Trans. Power Syst.* **1** 461-469